AF413165

A Gentle Introduction to Quantum Machine Learning

Yuxuan Du • Xinbiao Wang • Naixu Guo •
Zhan Yu • Yang Qian • Kaining Zhang •
Min-Hsiu Hsieh • Patrick Rebentrost • Dacheng Tao

A Gentle Introduction to Quantum Machine Learning

Springer

Yuxuan Du
College of Computing and Data Science
Nanyang Technological University
Singapore, Singapore

Naixu Guo
Centre for Quantum Technologies
National University of Singapore
Singapore, Singapore

Yang Qian
Department of Data Science
City University of Hong Kong
Hong Kong, China

Min-Hsiu Hsieh
Quantum Computing Research Center
Hon Hai (Foxconn)
Taipei, Taiwan

Dacheng Tao
College of Computing and Data Science
Nanyang Technological University
Singapore, Singapore

Xinbiao Wang
College of Computing and Data Science
Nanyang Technological University
Singapore, Singapore

Zhan Yu
Centre for Quantum Technologies
National University of Singapore
Singapore, Singapore

Kaining Zhang
College of Computing and Data Science
Nanyang Technological University
Singapore, Singapore

Patrick Rebentrost
Centre for Quantum Technologies
National University of Singapore
Singapore, Singapore

ISBN 978-981-95-1283-6   ISBN 978-981-95-1284-3   (eBook)
https://doi.org/10.1007/978-981-95-1284-3

This Springer imprint is published by the registered company Springer Nature Singapore Pte Ltd.
The registered company address is: 152 Beach Road, #21-01/04 Gateway East, Singapore 189721, Singapore

If disposing of this product, please recycle the paper.

Preface

Quantum computers, as next-generation computational devices, harness the quantum principles of superposition and entanglement to process information in ways fundamentally different from classical computers. These unique properties enable quantum computers to address many practical problems that are intractable for classical computers. Although quantum computing is still in its early stages, we have entered an era since 2019 where quantum supremacy has been experimentally demonstrated by several research groups and industrial organizations, underscoring the immense potential of quantum technologies to transform various aspects of everyday life.

Machine learning (ML) is widely regarded as one of the most promising and impactful applications of quantum computing. The ability of quantum computing to accelerate advancements in foundational models, such as generative pre-trained transformers (GPTs), and even pave the way toward artificial general intelligence (AGI), is particularly compelling. Recent progress in both theories and experiments has exhibited the power of quantum machine learning (QML). More precisely, the integration of quantum computing with ML may lead to novel approaches that outperform classical algorithms by offering faster runtimes, better performance, and reduced data requirements. This advancement can benefit many areas such as computer vision, natural language processing, drug discovery, finance, and fundamental science.

As an interdisciplinary field, the development of QML requires close collaboration between leading scientists and engineers in both quantum computing and artificial intelligence (AI). At the same time, as QML advances alongside the continuous progress of quantum hardware, there is a growing need for expertise from the AI community to drive this emerging field forward. However, the distinct conceptual frameworks and terminologies of quantum and classical computing present significant barriers for researchers and practitioners with a classical ML background in understanding the mechanisms behind QML algorithms and the benefits they may offer. Reducing this barrier to entry remains a major challenge within the community.

To overcome this challenge, we have written this book to deliver a comprehensive introduction to the latest developments in QML. Whether you are an AI researcher, an ML practitioner, or a computer science student, this resource will equip you with a solid foundation in the principles and techniques of QML. By bridging the gap between classical ML and quantum computing, this book could serve as a useful resource for those looking to engage with QML and explore the forefront of AI in the quantum era.

Singapore, Singapore   Yuxuan Du
Singapore, Singapore   Xinbiao Wang
Singapore, Singapore   Naixu Guo
Singapore, Singapore   Zhan Yu
Hong Kong, China   Yang Qian
Taipei, Taiwan   Min-Hsiu Hsieh
Singapore, Singapore   Patrick Rebentrost
Singapore, Singapore   Dacheng Tao
February 2025

Declarations

Competing Interests The authors have no competing interests to declare that are relevant to the content of this manuscript.

Contents

List of Acronyms

Notations Used in This Book

Notation	Concept
$a, b, \boldsymbol{a}_j, \boldsymbol{b}_j, \alpha, \beta$	Scalars
$\boldsymbol{x}, \boldsymbol{y}$	Vectors
W, A	Matrices
$\mathbb{R}$	Real Euclidean space
$\mathbb{C}$	Complex Euclidean space
$\mathbb{N}$	The set of natural numbers
$[a]$	The set of integers $\{1, 2, \cdots, a\}$
$\mathbb{E}[\cdot]$	Expectation value of a random variable
$\mathrm{Var}[\cdot]$	Variance of a random variable
O	Asymptotic upper bound notation
Ω	Asymptotic lower bound notation
$\top$	Transpose operation
$a*$	Complex conjugate of the number a
$\dagger$	Conjugate transpose operation
$\lvert \cdot \rangle \langle \cdot \rvert$	Outer product operation
$A \odot B$	Element-wise multiplication (Hadamard product) of matrices A and B
$f \circ g$	Composition of functions f and g
$\lvert 0 \rangle, \lvert 1 \rangle, \lvert \psi \rangle$	Pure quantum state in Dirac notation
N	Number of qubits
$\lvert 0^N \rangle, \lvert 0 \rangle^{\otimes N}$	Zero state with N-qubits
$\mathbb{I}_d$	Identity matrix with the size $d \times d$
ρ	Quantum state in density matrix representation
H	Hamiltonian
U, V	Unitary operator
X, Y, Z	Pauli operators
RX, RY, RZ	Rotation gates along the x, y, and z axes, respectively

CX, CZ	Controlled-X gate and controlled-Z gate
$\mathcal{E}, \mathcal{N}$	Quantum channel
O	Observable
$\langle O \rangle$	Expectation of observable O
$\|\cdot\|_{op}$	Operator norm
n	Number of training examples
$\mathcal{D}$	Dataset
$\mathrm{Tr}(O\rho)$	Expectation value of an observable O
$U(\boldsymbol{\theta})$	Parameterized quantum circuit
$\mathcal{L}(\boldsymbol{\theta})$	Loss function
$\nabla_{\boldsymbol{\theta}} \mathcal{L}(\boldsymbol{\theta})$	Gradient of loss function $\mathcal{L}$ w.r.t parameters $\boldsymbol{\theta}$

A Glossary of Key QML Terms

Term	Concept
NISQ	The terminology "Noisy Intermediate-Scale Quantum" (NISQ) is a summary of the current status of quantum computing hardware, characterized by devices with 50 to a few hundred qubits that are prone to noise and errors.
Ansatz / VQC / PQC	Ansatz, Variational Quantum Circuit (VQC), and Parameterized Quantum Circuit (PQC) are interchangeably used in literature to refer to a parameterized quantum state or a specific form of a quantum circuit used as a trial solution for variational algorithms.
HEA	Hardware Efficient Ansatz (HEA) is a design strategy for VQCs that minimizes circuit depth and gate count by using the native gate set and connectivity of current quantum hardware.
Quantum Dataset	A collection of data that describes quantum systems and their evolution.
Quantum Feature Map	A method used in QML to encode classical data into quantum states.
Basis Encoding	Directly encodes a classical binary string into quantum states, e.g., encoding 001 as $|001\rangle$.
Amplitude Encoding	Encodes a normalized classical vector into amplitudes of a quantum state, e.g., encoding a classical vector (x_1, x_2) as $x_1|0\rangle + x_2|1\rangle$.
Angle Encoding	Encodes data into angles of quantum rotation gates, e.g., a classical value x is mapped to a state using a rotation gate: $|\psi(x)\rangle = \mathrm{RY}(x)|0\rangle$.
VQA	Variational Quantum Algorithm (VQA) represents a class of hybrid quantum-classical algorithms that use

	VQCs/PQCs optimized by classical algorithms. Two key classes of VQAs are quantum neural networks and variational quantum eigen-solvers.
QNN	The term "Quantum Neural Network" (QNN) denotes a quantum-classical hybrid framework in which quantum circuits with adjustable parameters are optimized to perform tasks such as classification, regression, or simulation.
QCNN	Quantum Convolutional Neural Network (QCNN) is the extension of classical convolutional neural network in the quantum scenario. Two key components of QCNNs are quantum convolutional layers and pooling layers. The former is translationally invariant and the latter reduces the number of qubits while preserving key information.
Quantum Kernels	A similarity function that measures how close two classical data points are when mapped into a quantum Hilbert space.
Quantum Classifiers	Quantum machine learning models, including both quantum kernels and QNNs, are designed to classify data into different categories.
QGMs	Quantum Generative Models (QGMs) refer to a class of quantum machine learning models leveraging quantum computing to generate new data samples that follow the distribution of a given dataset. Common QGMs include quantum circuit Born machine, quantum generative adversarial network, quantum Boltzmann machine, quantum variational autoencoder, and quantum generative diffusion model.
Quantum Gradient	The derivative of a quantum computing function with respect to the parameters of a parameterized quantum circuit.
Parameter Shift Rule	A technique for computing quantum gradients by evaluated parameter-shifted instances of a variational quantum circuit.
BP	Barren Plateau (BP) is a phenomenon in the optimization landscape of certain VQAs where the gradient of the cost function becomes exponentially small as the number of qubits increases.
Expressivity	The capability of a quantum model or circuit (e.g., an ansatz) to represent a wide variety of quantum states or functions. In other words, it reflects how rich the model's representational power is. Common metrics of the expressivity of a QML model include Rademacher complexity, Vapnik–Chervonenkis (VC) dimension, covering number, and the size of the Fourier spectrum.

Trainability	A measure of how feasibly a quantum model can be optimized or trained using classical or hybrid optimization methods. It focuses on the structure of the quantum circuit's parameter space. Poor trainability, often due to phenomena like barren plateaus (i.e., regions where gradients vanish), can significantly hinder the learning process.
Generalization error	The difference in performance between a quantum model's predictions on training data and its performance on new, unseen data. In QML, a low generalization error indicates that the model has effectively captured the underlying patterns without overfitting the training set, ensuring it can robustly handle novel inputs.
FTQC	Fault-tolerant quantum computing (FTQC) is a system that performs reliable quantum computations despite the presence of errors and noise in quantum hardware.
Quantum Noise	Unwanted interactions between a quantum system and its environment, leading to decoherence and computational errors.
QEC	Quantum error correction (QEC) is a set of techniques designed to protect quantum information from errors caused by decoherence, gate imperfections, and measurement noise.
QRAM	Quantum random access memory (QRAM) is a quantum data structure that allows efficient access to large datasets in superposition.
Block Encoding	A quantum algorithmic primitive for embedding arbitrary matrices into unitary matrices that can be implemented on a quantum computer.
QSP	Quantum signal processing (QSP) is a framework that can implement nonlinear polynomials of signals by iterating signal unitaries and signal processing unitaries.
QSVT	Quantum singular value transformation (QSVT) is the multi-qubit lifted version of QSP, which is a powerful technique for applying polynomial transformations to singular values of a matrix.
LCU	Linear combination of unitaries (LCU) is a technique that implements the block encoding of a weighted sum of unitary matrices, enabling efficient quantum matrix operations.

Chapter 1
Introduction

Abstract This chapter introduces the emerging field of quantum machine learning (QML), which aims to integrate the strengths of quantum computing with classical machine learning to achieve computational advantages. It begins by contextualizing QML within the broader evolution of computing paradigms, from CPUs and GPUs to quantum processors, and highlights the limitations of classical systems in handling increasingly complex learning tasks. The chapter then presents a first glimpse into QML, including the foundational components of quantum computers, different measures of quantum advantages, and the main research directions in QML. Progress in QML is reviewed under two regimes: fault-tolerant quantum computing (FTQC) and noisy intermediate-scale quantum (NISQ) devices. Finally, the chapter outlines the structure of the tutorial and explains its intended audience. Together, these discussions provide a comprehensive background and motivation for the chapters that follow.

The advancement of computational power has been a central driver of modern industrial revolutions, particularly since the mid-twentieth century. The invention of the modern computer, followed by the central processing unit (CPU), led to the "digital revolution," transforming industries through process automation and the rise of information technology. Later, the emergence of graphical processing units (GPUs) accelerated advancements in artificial intelligence (AI) and big data, making applications such as intelligent transportation, autonomous vehicles, scientific simulations, and data analysis possible. Moore's law, which describes the doubling of transistors on integrated circuits every two years, is reaching its physical and practical limits. Traditional computing hardware, including CPUs and GPUs, is constrained by these limitations. The exponential growth of data and increasing complexity of applications require new computational paradigms. **Quantum computing** [1] emerges as a promising approach by harnessing principles of quantum mechanics, such as superposition and entanglement, to process information in fundamentally new ways.

One of the most concrete and direct ways to understand the potential of quantum computers is through the framework of complexity theory [2]. Theoretical computer

Y. Du et al., *A Gentle Introduction to Quantum Machine Learning*,
https://doi.org/10.1007/978-981-95-1284-3_1

scientists have demonstrated that quantum computers can efficiently solve problems within the BQP (bounded-error quantum polynomial time) complexity class, meaning these problems can be solved in polynomial time by a quantum computer. In contrast, classical computers are limited to efficiently solving problems within the P (polynomial time) complexity class. While it is widely believed, though not proven, that $P \subseteq BQP$, this suggests that quantum computers can provide exponential speedups for certain problems in BQP that are intractable for classical machines.

A prominent example of such a problem is large-number factorization, which forms the basis of RSA cryptography. Shor's algorithm [3], a quantum algorithm, can factor large numbers in polynomial time, while the most efficient known classical factoring algorithm requires super-polynomial time. For instance, breaking an RSA-2048 bit encryption key would take a classical computer approximately 300 trillion years, whereas an *ideal* quantum computer could complete the task in around 10 seconds. However, constructing "ideal" quantum computers remains a significant challenge. As will be discussed in later chapters, based on current fabrication techniques, this task could potentially be completed in approximately 8 hours using a noisy quantum computer with a sufficient number of **qubits**—the fundamental units of quantum computation [4].

The convergence of the computational power offered by quantum machines and the limitations faced by AI models has led to the rapid emergence of the field: **quantum machine learning** (QML) [5]. In particular, the challenges in modern AI stem from the neural scaling law [6], which posits that "bigger is often better." Since 2020, this principle has driven the development of increasingly colossal models, featuring more complex architectures and an ever-growing number of parameters. However, this progress comes at an immense cost. For instance, training a model like ChatGPT on a single GPU would take approximately 355 years, while the cloud computing costs for training such large models can reach tens of thousands of dollars.

These staggering costs present a critical barrier to the future growth of AI. Quantum computing, celebrated for its extraordinary computational capabilities, holds the potential to overcome these limitations. It offers the possibility of advancing models like generative pretrained transformers (GPTs) and accelerating progress toward artificial general intelligence (AGI). Quantum computing, particularly QML, marks a shift from the classical "it from bit" paradigm to the quantum "it from qubit" perspective. This transition has the potential to reshape AI and computational science.

1.1 Introduction to Quantum Machine Learning

What exactly is quantum machine learning (QML)? In its *simplest* terms, the focus of this book on QML can be summarized as follows (see Sect. 1.1.3 for the systematic overview).

> **Quantum Machine Learning (Informal)**
> QML explores learning algorithms that can be executed on quantum computers to accomplish specified tasks with potential advantages over classical implementations.

This interpretation involves three key elements: *quantum processors*, *specified tasks*, and *advantages*. The following sections clarify the specific meaning of each term, laying the groundwork for understanding the mechanisms and potential of QML.

1.1.1 Quantum Computers

Quantum computing traces its origins to 1980, when Paul Benioff proposed the *quantum Turing machine* [7], a quantum analog of the classical Turing machine. Several models of quantum computation have since emerged, including *circuit-based quantum computation* [8], *one-way quantum computation* [9], *adiabatic quantum computation* [10], and *topological quantum computation* [11]. These models are computationally equivalent, as any can efficiently simulate the others. Due to its prevalence in both research and industry, the **circuit-based quantum computer** is the primary focus in this book.

Quantum computing gained further momentum in the early 1980s when physicists faced an exponential increase in computational overhead while simulating quantum dynamics, particularly as the number of particles in a system grew. This "curse of dimensionality" prompted Yuri Manin and Richard Feynman to independently propose leveraging quantum phenomena to build quantum computers. Accordingly, such devices would be far more efficient for simulating quantum systems than classical computers.

However, as a universal computing device, the potential of quantum computers extends well beyond quantum simulations. In the 1990s, [3] developed a groundbreaking quantum algorithm for large-number factorization, posing a serious threat to widely used encryption protocols such as RSA and Diffie–Hellman. In 1996, Grover's algorithm demonstrated a quadratic speedup for unstructured search problems [12], a task with broad applications. Since then, the influence of quantum computing has expanded into a wide range of fields. To name a few, various quantum algorithms have been developed to achieve runtime speedups in finance [13], drug design [14], optimization [15], and, most relevant to this book, machine learning.

A direct comparison of fundamental components clarifies why quantum computers may outperform classical computers. As shown in Fig. 1.1, both classical and quantum computers feature three primary components: input, computation, and output. Table 1.1 summarizes the differences in their implementations.

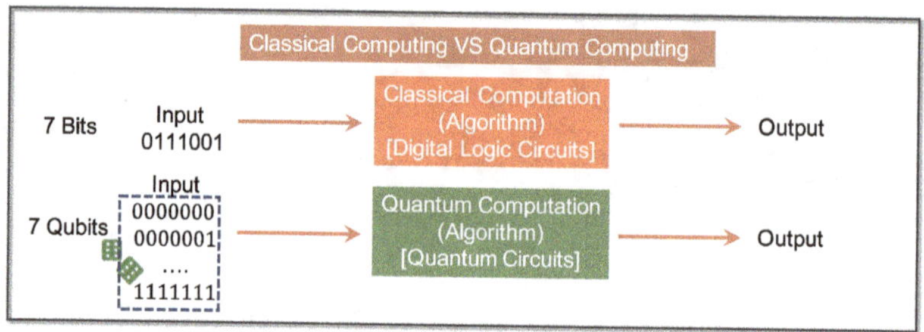

Fig. 1.1 The paradigm between classical and quantum computing

Table 1.1 Comparison between classical and quantum computing

	Classical	Quantum
Input	Binary bits	Quantum bits
Computation	Digital logical circuits	Quantum circuits
Output	Retrieve solution	Quantum measurements

Table 1.2 Mathematical representations of N-(quantum) bits in classical and quantum computers. Here, the symbols '$\dagger$' and $\mathbb{C}$ denote the transpose conjugation and complex space, respectively

	Classical	Quantum
Single bit $(N = 1)$	$x \in \{0, 1\}$	$[a_1, a_2]^{\dagger} \in \mathbb{C}^2$ s.t. $|a_1|^2 + |a_2|^2 = 1$
Multiple bits $(N > 1)$	$x \in \{0, 1\}^N$	s.t.$[a_1, a_2, \ldots, a_{2^N}]^{\dagger} \in \mathbb{C}^{2^N}$ s.t. $|a_1|^2 + |a_2|^2 + \ldots + |a_{2^N}|^2 = 1$

The advantages of quantum computers stem primarily from the key distinctions between classical bits and quantum bits (**qubits**), as well as between digital logic circuits and **quantum circuits**, as outlined below:

- *Bits versus qubits*. A classical bit is a binary unit that takes on a value of either 0 or 1. In contrast, a quantum bit, or qubit, can exist in a superposition of both 0 and 1 simultaneously, represented by a *two-dimensional vector* where the entries correspond to the *probabilities* of the qubit being in each state.

  Furthermore, while classical bits follow the Cartesian product rule, qubits adhere to the tensor product rule. This distinction implies that an N-qubit system is described by a 2^N-dimensional vector, allowing quantum systems to encode information exponentially with N—far surpassing the capacity of classical bits. Table 1.2 summarizes the mathematical expressions of classical and quantum bits.

- *Digital logic circuits versus quantum circuits*. Classical computers rely on digital logic circuits composed of logic gates that perform operations on bits in a deterministic manner, as illustrated in Fig. 1.1. In contrast, quantum circuits consist of **quantum gates**, which act on single or multiple qubits to modify their states—the probability amplitudes $a_1, \ldots, a_{2^N}$, as shown in Table 1.2. Owing to the universality of quantum gates, for any given input qubit state, there always exists a specific quantum circuit capable of transforming the input state into one corresponding to the target solution—a particular probability distribution. For certain probability distributions, a quantum computer can use a polynomial number of quantum gates relative to the qubit count N to generate the distribution. In contrast, classical computers require an exponential number of gates with N to achieve the same result. This difference underpins the quantum advantage.

- The readout process in quantum computing differs fundamentally from that in classical computing, as it involves **quantum measurements**. Intuitively, quantum measurements can extract information from a quantum system and translate it into a form that can be interpreted by classical systems. For problems in quantum physics and chemistry, quantum measurements can reveal far more useful information than classical simulations of the same systems, enabling significant runtime speedups in obtaining the desired physical properties.

The formal definitions of quantum computing are presented in Chap. 2. The computational power of quantum computers is primarily determined by two factors: the number of qubits and quantum gates as well as their respective qualities. The term "qualities" refers to the fact that fabricating quantum computers is highly challenging, as both qubits and quantum gates are prone to errors. These qualities are measured using various physical metrics. One commonly used metric is *quantum volume* V_Q [16], which quantifies a quantum computer's capabilities by accounting for both its error rates and overall performance. Mathematically, the quantum volume represents the maximum size of square quantum circuits that the computer can successfully implement to achieve the *heavy output generation problem*. The mathematical expression is

$$\log_2(V_Q) = \arg\max_m \min(m, d(m)), \tag{1.1}$$

where $m \leq N$ is a number of qubits selected from the given N-qubit quantum computer and $d(m)$ is the number of qubits in the largest square circuits for which heavy outputs can be reliably sampled with probability exceeding 2/3. The heavy output generation problem discussed here stems from proposals aimed at demonstrating quantum advantage. That is, a quantum computer of sufficiently high quality is expected to generate heavy outputs frequently across a range of random quantum circuit families. For illustration, Table 1.3 summarizes the progress of quantum computers as of 2024.

Table 1.3 Progress of quantum computers up to December 2024

Date	$\log_2(V_Q)$	N	Manufacturer	System name
Dec, 2024	–	105	Google	Willow
Aug, 2024	21	56	Quantinuum	H2-1
Jul, 2024	9	156	IBM	Heron
Jun, 2023	19	20	Quantinuum	H1-1
Sep, 2022	13	20	Quantinuum	H1-1
Apr, 2022	12	12	Quantinuum	H1-2
Jul, 2021	10	10	Honeywell	H1
Nov, 2020	7	10	Honeywell	H1
Aug, 2020	6	27	IBM	Falcon r4

Remark

Note that quantum volume is not the unique metric for evaluating the performance of quantum computers. There are several other metrics that assess the power of quantum processors from different perspectives. For instance, *circuit layer operations per second* (CLOPS) [17] measures the computing speed of quantum computers, reflecting the feasibility of running practical calculations that involve a large number of quantum circuits. Additionally, *effective quantum volume* [18] provides a more nuanced comparison between noisy quantum processors and classical computers, considering factors such as error rates and noise levels. These metrics, among others, offer a more comprehensive understanding of the strengths and limitations of quantum computers across various applications.

1.1.2 Metrics for Quantum Advantages

Quantum advantage refers to situations where quantum computers solve problems more efficiently than classical computers. However, "efficiency" can be defined in multiple ways. The most common metric is *runtime complexity*, where quantum algorithms may achieve significant or even exponential speedups. For example, Shor's algorithm provides exponential improvements for large-number factorization.

In the context of quantum learning theory [19], efficiency is often measured by *sample complexity*, especially within the probably approximately correct (PAC) learning framework. Here, sample complexity is defined as the number of interactions (e.g., queries of target quantum systems or measurements) required for a learner to achieve a desired prediction accuracy below a specified threshold. Here, the quantum advantage is realized when the upper bound on the sample complexity of a quantum learning algorithm for a given task is *lower than* the lower bound of all classical learning algorithms. While low sample complexity is a necessary condition for efficient learning, it does not guarantee practical efficiency alone. For example, identifying useful training examples within a small sample size may still require substantial computational time.

> **Remark (Difference of Sample Complexity in Classical and Quantum ML)**
>
> In classical ML, sample complexity typically refers to the number of training examples required for a model to generalize effectively. An example is the number of labeled images needed to train an image classifier. In quantum ML, however, sample complexity can take on varied meanings depending on the context, as shown below.
>
> - Quantum state tomography (see Sect. 2.3.2). Here, the sample complexity refers to the number of measurements required to accurately reconstruct the quantum state of a system.
> - Evaluation of the generalization ability of quantum neural networks (see Sect. 4.4). Here, the sample complexity refers to the number of input-output pairs needed to train the network to approximate a target function, similar to classical ML.
> - Quantum system learning. Here, the sample complexity often refers to the number of queries to interact with the target quantum system, such as the number of times a system must be probed to learn its Hamiltonian dynamics.

In addition to sample complexity, another commonly used measure in quantum learning theory is *quantum query complexity*, particularly within the frameworks of quantum statistical learning and quantum exact learning. As these frameworks are not the primary focus of this book, interested readers are referred to [20] for a more detailed discussion.

Quantum advantage can be pursued through *two main approaches*. The first involves identifying problems with quantum circuits that demonstrate provable advantages over classical counterparts in the aforementioned measures [21]. Such findings deepen our understanding of quantum computing's potential and expand its range of applications. However, these quantum circuits often require substantial quantum resources, which are currently beyond the reach of near-term quantum computers. Additionally, for many tasks, analytically determining the upper bound of classical algorithm complexities is challenging.

These challenges have motivated a second approach: demonstrating that current quantum devices can perform accurate computations on a scale that exceeds brute-force classical simulations—a milestone known as "quantum utility." Quantum utility refers to quantum computations that yield reliable, accurate solutions to problems beyond the reach of brute-force classical methods and otherwise accessible only through classical approximation techniques [22]. This approach represents a step toward practical computational advantage with noise-limited quantum circuits. Reaching the era of quantum utility signifies that quantum computers have attained a level of scale and reliability enabling researchers to use them as effective tools for scientific exploration, potentially leading to groundbreaking new insights.

1.1.3   Explored Tasks in Quantum Machine Learning

QML research is extensive and can be broadly categorized into four main areas. These categories are defined by two factors: the nature of the computing device (whether it is quantum (Q) or classical (C)) and the type of data being processed (whether it is generated by a quantum (**Q**) or classical (**C**)) system) and the type of data being processed (whether it is generated by a quantum (**Q**) or classical (**C**) system). The four sectors are explained as follows:

CC Sector. The **CC** sector refers to classical data processed on classical systems, representing traditional machine learning. Here, classical ML algorithms run on classical processors (e.g., CPUs and GPUs) and are applied to classical datasets. A typical example is using neural networks to classify images of cats and dogs.

CQ Sector. The **CQ** sector involves using classical ML algorithms on classical processors to analyze quantum data collected from quantum systems. Typical examples include applying classical neural networks to classify quantum states, estimating properties of quantum systems from measurement data, and employing classical regression models to predict outcomes of quantum experiments.

QC Sector. The **QC** sector involves developing QML algorithms that run on quantum processors (QPUs) to process classical data. In this context, quantum computing resources are leveraged to enhance or accelerate the analysis of classical datasets. Typical examples include applying QML algorithms, such as quantum neural networks and quantum kernels, to improve pattern recognition in image analysis.

QQ Sector. The **QQ** sector involves developing QML algorithms executed on QPUs to process quantum data. In this context, quantum computing resources are leveraged to reduce the computational cost of analyzing and understanding complex quantum systems. Typical examples include using quantum neural networks for quantum state classification and applying quantum-enhanced algorithms to simulate quantum many-body systems.

The classification above is not exhaustive. As illustrated in Fig. 1.2, each sector can be further subdivided based on various learning paradigms. For instance, discriminative vs. generative learning or supervised, unsupervised, and semi-supervised learning. Additionally, each sector can be further categorized according to different application domains, such as chemistry, computer vision, power systems, logistics, finance, and healthcare.

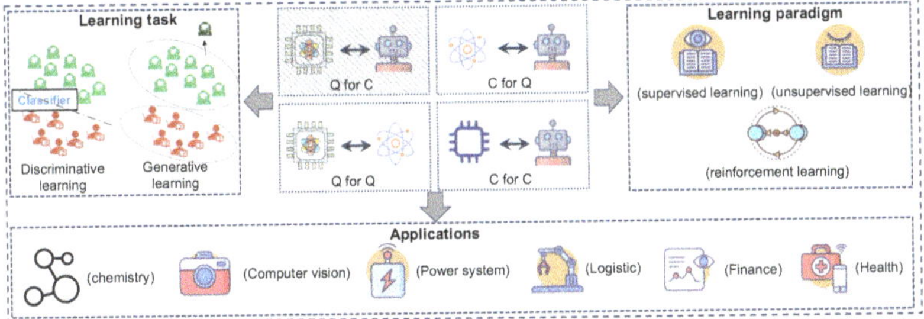

Fig. 1.2 Different research directions in QML

> **Remark**
> The primary focus of this book is on the **QC** and **QQ** sectors. For more details on **CQ**, interested readers can refer to [23–25].

1.2 Progress of Quantum Machine Learning

Significant efforts have been directed toward the **QC** and **QQ** sectors to identify the tasks and conditions under which QML can achieve computational advantages over classical machine learning. To clarify the progress of QML, it is essential to first examine the recent advances in quantum computers—the foundational infrastructure underpinning quantum algorithms.

1.2.1 Progress of Quantum Computers

The novelty and inherent challenges of utilizing quantum physics for computation have driven the development of various computational architectures, giving rise to the formalized concept of circuit-based quantum computers, as discussed in Sect. 1.1.1. In pursuit of this goal, numerous companies and organizations are striving to establish their architecture as the leading approach and to be the first to demonstrate practical utility or quantum advantage on a large-scale quantum device.

Common architectures currently include superconducting qubits (employed by IBM and Google), ion-trap systems (pioneered by IonQ), and Rydberg atom systems (developed by QuEra), each offering distinct advantages [26]. Specifically, superconducting qubits excel in scalability and fast gate operations [27], while ion-trap systems are known for their high coherence times, precise control over individual qubits, and full connectivity of all qubits [28]. Moreover, Rydberg atom

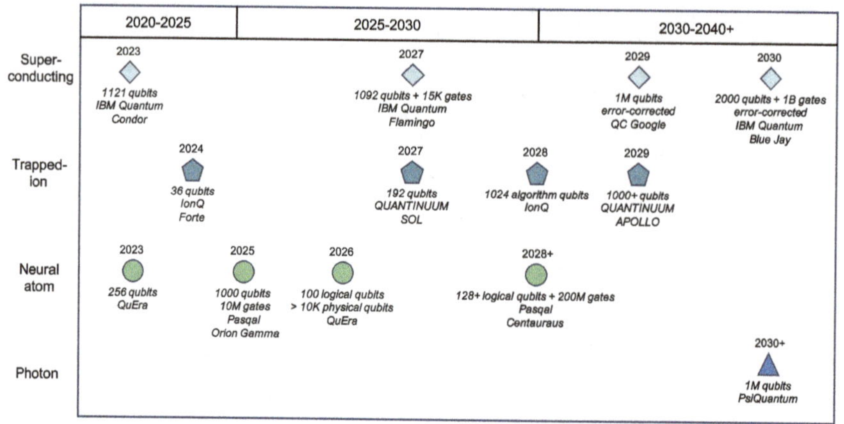

Fig. 1.3 Common quantum architectures and roadmaps from different quantum companies

systems enable flexible qubit connectivity through highly controllable interactions [29]. Besides these architectures, integrated photonic quantum computers are emerging as promising alternatives for robust and scalable quantum computation. A summarization is provided in Fig. 1.3.

Despite recent advances, today's quantum computers remain highly sensitive to environmental noise and prone to quantum decoherence, lacking the stability needed for fault-tolerant operation. This results in qubits, quantum gates, and quantum measurements that are inherently imperfect, introducing errors that can lead to incorrect outputs. To capture this stage in quantum computing, John Preskill coined the term "noisy intermediate-scale quantum" (NISQ) era [30], which describes the current generation of quantum processors. These processors feature up to thousands of qubits, but their capabilities are restricted with error-prone gates and limited coherence times.

In the NISQ era, notable achievements have been made alongside new challenges. Industrial and academic teams, such as those at Google and USTC, have demonstrated quantum advantages on specific sampling tasks, where the noisy quantum computers they fabricated outperform classical computers in computational efficiency [31, 32]. However, most quantum algorithms that theoretically offer substantial runtime speedups depend on fault-tolerant, error-free quantum systems—capabilities that remain beyond the reach of current technology.

At this pivotal stage, the path forward in quantum computing calls for progress on both hardware and algorithmic fronts.

On the hardware side, it is essential to continuously improve qubit count, coherence times, gate fidelities, and the accuracy of quantum measurements across various quantum architectures. Once the number and quality of qubits surpass certain thresholds, *quantum error correction* codes can be implemented [8], paving the way for fault-tolerant quantum computing (FTQC). Broadly, quantum error correction uses redundancy and entanglement to detect and correct errors without directly measuring the quantum state, thus preserving coherence. Advancements in quantum processors will enable a progression from the NISQ era to the early FTQC era, ultimately reaching the fully FTQC era.

On the algorithmic side, two key questions must be addressed:

- How can NISQ devices be utilized to perform meaningful computations with practical utility?
- What types of quantum algorithms can be executed on early fault-tolerant and fully fault-tolerant quantum computers to realize the potential of quantum computing in real-world applications?

Progress on either question could have broad implications. A positive answer to the first question would suggest that NISQ quantum computers have immediate practical applicability. Meanwhile, the advancements in the second question would expand the scope and impact of quantum computing as more robust, fault-tolerant systems become feasible. The following two sections review recent progress in quantum machine learning related to these two questions.

1.2.2 Progress of Quantum Machine Learning Under FTQC

A key milestone in FTQC-based QML algorithms is the quantum linear equations solver introduced by Harrow et al. [33]. Many machine learning models rely on solving linear equations, a computationally intensive task that often dominates the overall runtime due to the polynomial scaling of complexity with matrix size. The HHL algorithm provides a breakthrough by reducing runtime complexity to polylogarithmic scaling with matrix size, given that the matrix is well conditioned and sparse. This advancement is highly significant for AI, where datasets frequently reach sizes in the millions or even billions.

The exponential runtime speedup achieved by the HHL algorithm has garnered significant attention from the research community, highlighting the potential of quantum computing in AI. Following this milestone, a body of work has emerged that employs the quantum matrix inversion techniques developed in HHL (or its variants) as subroutines in the design of various FTQC-based QML algorithms. These algorithms often offer runtime speedups over their classical counterparts [34, 35]. Notable examples include quantum principal component analysis [36] and quantum support vector machines [37].

Another milestone in FTQC-based QML algorithms is the quantum singular value transformation (QSVT), proposed by Gilyén et al. [38]. QSVT enables polynomial transformations of the singular values of a linear operator embedded within a unitary matrix, offering a unifying framework for various quantum algorithms. It has connected and enhanced a broad range of quantum techniques, including amplitude amplification, quantum linear system solvers, and quantum simulation methods. Compared to the HHL algorithm for solving linear equations, QSVT provides improved scaling factors, making it a more efficient tool for addressing these problems in the context of QML.

In addition to advancements in linear equation solving, another promising line of research in FTQC-based QML focuses on leveraging quantum computing to enhance deep neural networks (DNNs) rather than traditional machine learning models. This research track has two main areas of focus. The first is the acceleration of DNN optimization, with notable examples including the development of efficient quantum algorithms for dissipative differential equations to expedite (stochastic) gradient descent, as well as quantum Langevin dynamics for optimization [39, 40]. The second area centers on advancing Transformers using quantum computing. In Chap. 5, how quantum computing can be employed to accelerate Transformers during the inference stage will be discussed in detail.

Remark

However, there are several critical caveats of the HHL-based QML algorithms. First, the assumption of efficiently preparing the quantum states corresponding to classical data runtime is very strong and may be impractical in the dense setting. Second, the obtained result x is still in the quantum form $|x\rangle$. Note that extracting one entry of $|x\rangle$ into the classical form requires $O(\sqrt{N})$ runtime, which collapses the claimed exponential speedups. The above two issues amount to the read-in and read-out bottlenecks in QML [41]. The last caveat is that the employed strong quantum input model such as quantum random access memory (QRAM) [42] leads to an inconclusive comparison. Through exploiting a classical analog of QRAM as the input model, there exist efficient classical algorithms to solve recommendation systems in polylogarithmic time in the size of input data.

1.2.3 Progress of Quantum Machine Learning Under NISQ

The work conducted by Havlicek et al. [43] marked a pivotal moment for QML in the NISQ era. This study demonstrated the implementation of quantum kernel methods and quantum neural networks (QNNs) on a 5-qubit superconducting quantum computer, highlighting potential quantum advantages from the perspective of complexity theory. Unlike the aforementioned FTQC algorithms, quantum kernel methods and QNNs are flexible and can be effectively adapted to the limited quantum resources available in the NISQ era. These demonstrations, along with advancements in quantum hardware, sparked significant interest in exploring QML applications using NISQ quantum devices. We will delve into quantum kernel methods and QNNs in Chaps. 3 and 4, respectively.

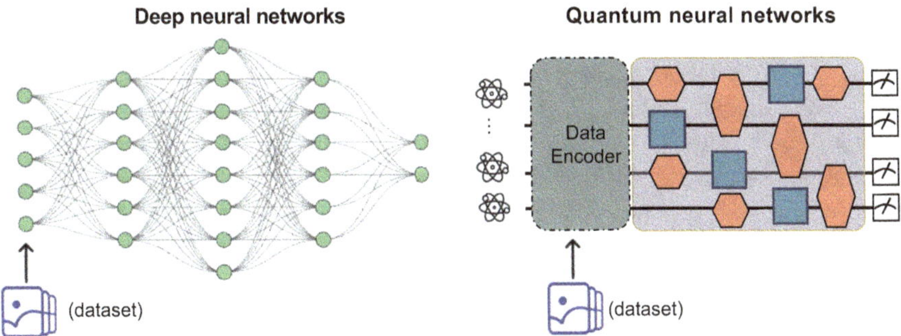

Fig. 1.4 Mechanisms of DNNs and QNNs

> **Quantum Neural Networks (Informal)**
> A quantum neural network (QNN) is a hybrid model that leverages quantum
> computers to implement trainable models similar to classical neural networks
> while using classical optimizers to complete the training process.

As shown in Fig. 1.4, the mechanisms of QNNs and deep neural networks
(DNNs) are almost the same. Both DNNs and QNNs follow an iterative approach.
At each iteration, they take input data, process it through multiple layers, and
produce an output prediction. The key difference between DNNs and QNNs is the
way of implementing their learning models. This difference gives the potential of
quantum learning models to solve complex problems beyond the reach of classical
neural networks, opening new frontiers in many fields. Roughly speaking, research
in QNNs and quantum kernel methods has primarily focused on three key areas:
(I) *quantum learning models and applications*, (II) *the adaptation of advanced AI
topics to QML*, and (III) *theoretical foundations of quantum learning models*. A
brief overview of each category is provided below.

(I) QUANTUM LEARNING MODELS AND APPLICATIONS. This category focuses
on implementing various DNNs on NISQ quantum computers to tackle a wide range
of tasks.

From a model architecture perspective, quantum analogs of popular classical
machine learning models have been developed. Typical instances include the quantum versions of multilayer perceptrons (MLPs), autoencoders, convolutional neural
networks (CNNs), recurrent neural networks (RNNs), extreme learning machines,
generative adversarial networks (GANs), diffusion models, and Transformers. Some
of these QNN structures have even been validated on real quantum platforms,
demonstrating the feasibility of applying quantum algorithms to tasks traditionally
dominated by classical deep learning [44–46].

From an application perspective, QML models implemented on NISQ devices
have been explored across diverse fields, including fundamental science, image

classification, image generation, financial time series prediction, combinatorial optimization, healthcare, logistics, and recommendation systems. These applications demonstrate the broad potential of QML in the NISQ era, though achieving full quantum advantage in these areas remains an ongoing challenge [47, 48].

(II) ADAPTATION OF ADVANCED AI TOPICS TO QML. Beyond model design, advanced topics from AI have been extended to QML, aiming to enhance the performance and robustness of different QML models. Examples include quantum architecture search [49] (the quantum equivalent of neural architecture search), advanced optimization techniques [50], and pruning methods to reduce the complexity of quantum models [51, 52]. Other areas of active research include adversarial learning [53], continual learning [54, 55], differential privacy [56, 57], distributed learning [58, 59], federated learning [60], and interpretability within the context of QML [61]. These techniques have the potential to significantly improve the efficiency and effectiveness of QML models, addressing some of the current limitations of NISQ devices.

(III) THEORETICAL FOUNDATIONS. Quantum learning theory [62] has garnered increasing attention, aiming to compare the capabilities of different QML models and to identify the theoretical advantages of QML over classical machine learning models. As shown in Fig. 1.5, the learnability of QML models can be evaluated across three key dimensions: expressivity, trainability, and generalization capabilities. Similar to classical learning theory, these dimensions serve as the foundation for understanding how quantum models perform in various contexts, providing a common ground for comparing QML and classical ML models. Below, a brief overview of each measure is provided.

- Trainability. This area examines how the design of QNNs influences their convergence properties, including the impact of system noise and measurement errors on the ability to converge to local or global minima. Good trainability allows QML models to efficiently optimize toward an optimal solution. Due to its importance, we address two key concepts with respect to the trainability of QNNs, i.e., barren plateaus and overparameterization, in Sect. 4.4.2.

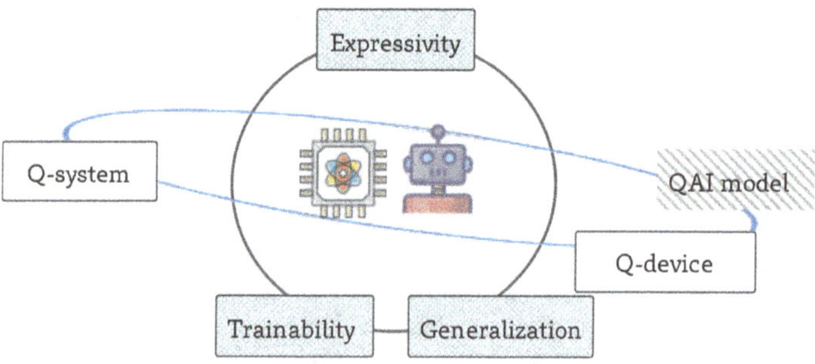

Fig. 1.5 The learnability of quantum machine learning models

- Expressivity. Researchers investigate how the number of parameters and the structure of QNNs affect the size of the hypothesis space they can represent. A central question is whether QNNs and quantum kernels can efficiently represent functions or patterns that classical neural networks cannot, thereby offering potential quantum advantage. In Sects. 3.3.1 and 4.4.1, the expressivity of QML models is explored from the perspectives of the universal approximation theorem and model complexity, respectively.
- Generalization. This focuses on understanding how the gap between training and test error evolves with the size of the dataset, the structure of QNNs or quantum kernels, and the number of parameters. The goal is to determine whether QML models can generalize more effectively than classical models, particularly in the presence of noisy data or when training data is limited. For comprehensive, Sects. 3.3.2 and 4.4.1, an in-depth analysis of how QML models generalize with limited training data and varying model structures is provided, along with the interpretation of the generalization advantages of QML models for specific datasets.

The combination of advancements in model design, application domains, and theoretical understanding is driving the progress of QML in the NISQ era. Although the field is still in its early stages, the progress achieved thus far provides promising insights into the potential of quantum computing to enhance conventional AI. As quantum hardware continues to evolve, further breakthroughs are expected, potentially unlocking new possibilities for practical QML applications.

> **Remark**
> It is important to note that QNNs and quantum kernel methods can also be considered FTQC algorithms when executed on fully fault-tolerant quantum computers. The reason these algorithms are discussed in the context of NISQ devices is their flexibility and robustness, making them well suited to the limitations of current quantum hardware.

Unlike quantum hardware, where the number of qubits has rapidly scaled from zero to thousands, the development of QML algorithms—and quantum algorithms more broadly—has taken an inverse trajectory, transitioning from FTQC to NISQ devices. This shift reflects the move from idealized theoretical frameworks to practical implementations. The convergence of quantum hardware and QML algorithms, where the quantum resources required by these algorithms become attainable on real quantum computers, enables researchers to experimentally evaluate the power and limitations of various quantum algorithms.

Based on the minimum quantum resources required to complete learning tasks, we distinguish between FTQC algorithms, discussed in Sect. 1.2.2, and NISQ algorithms, including QNNs and quantum kernel methods, in Sect. 1.2.3. FTQC-based QML algorithms necessitate error-corrected quantum computers with tens of billions of qubits—an achievement that remains far from realization. In contrast, QNNs and quantum kernels are more flexible and can be executed on both NISQ and FTQC devices, depending on the available resources.

As quantum hardware continues to progress, the development of QML algorithms must evolve in tandem. A promising direction is to integrate FTQC algorithms with QNNs and quantum kernel methods, creating new QML algorithms that can be run on current quantum processors while offering enhanced quantum advantages across various tasks.

To maintain self-consistency, this section concludes by highlighting key differences between classical and quantum machine learning models in complexity, representational power, and scalability. QML models can achieve exponential speedups for specific computational tasks, such as large-number factorization. They also provide enhanced representational power through superposition and entanglement, supporting more compact and expressive data representations. Furthermore, QML models have the potential for greater scalability compared to classical approaches, although practical scalability is currently limited by hardware and error correction constraints. These advantages underscore the motivation for continued exploration of QML.

1.3 Organization of This Book

To encourage and enable computer scientists to engage with the rapidly growing field of quantum AI, we provide this book that revisits QML algorithms from a *computer science perspective*. With this aim, the book is designed to balance theory, practical implementations, and applications, making it suitable for readers with some background in classical machine learning. The book is divided into the following chapters:

Chapter 2: BASICS OF QUANTUM COMPUTING. Before delving into QML, this chapter lays the groundwork by introducing the fundamental concepts of quantum computing. It covers the transition from classical bits to quantum bits, explains quantum circuit models, illustrates how quantum systems interface with classical systems through quantum read-in and read-out mechanisms, and presents some fundamental concepts of quantum linear algebra. By the end of this chapter, you will understand that a solid grasp of linear algebra is all you need to comprehend the basics of quantum computing.

Chapters 3, 4, and 5: CLASSICAL ML MODELS EXTENDED TO QUANTUM FRAMEWORKS. Each of these chapters follows a consistent structure, starting with a review of the classical model and progressing to its quantum extension— quantum kernel methods in Chap. 3, quantum neural networks in Chap. 4, and quantum Transformers in Chap. 5. This unified structure enables readers to clearly understand how classical machine learning models can be translated into quantum implementations and how quantum computers may offer computational advantages.

Appendix
The appendix serves as a supplementary resource, providing the essential mathematical tools that are omitted from the main text for brevity. In particular, it includes basic introduction of concentration inequalities, the Haar measure, and other foundational concepts relevant to the book.

To provide a clear and comprehensive learning experience, each chapter is composed of the following parts:

1. Classical foundations and quantum model construction. Each chapter begins with a review of the classical version of the model, ensuring that readers are well acquainted with the foundational concepts before exploring their quantum adaptations. After this review, we introduce quantum versions of the models, focusing on implementations based on NISQ, FTQC, or both.

2. Theoretical analysis. There is nothing more practical than a good theory. In this book, each chapter provides a theoretical analysis of the learnability of QML models, focusing on key aspects such as expressivity, trainability, and generalization capabilities.

   To ensure a balance between depth and self-consistency, this book provides proof for the most significant theoretical results, as highlighted by **theorems and lemmas**. For results that are less central to the main content of this book, we present them as **facts** and include appropriate references, allowing readers to easily locate the complete proofs if desired.

3. Code implementation. "Talk is cheap, show me the code." To provide a practical, hands-on learning experience, each chapter includes code implementations using real-world datasets (to be specified). This section walks readers through the process of implementing quantum models on simulated or real quantum hardware. All numerical examples illustrated in this book are available in https://qml-tutorial.github.io/, accompanied by Jupyter Notebooks.

   Instead of building everything from scratch, the well-established PennyLane library is employed for implementation [63]. This choice does not imply any specific preference. Other quantum computing libraries, such as Qiskit [64], Cirq [65], and TensorFlow Quantum [66], can also be used, offering similar capabilities and flexibility. A brief summary of these quantum computing tools is listed in Table 1.4.

4. Frontier topics and future directions. Each chapter concludes with an exploration of cutting-edge topics and emerging challenges in the field. This part highlights open research problems, ongoing developments, and potential future directions for the quantum versions of each model, providing insights into where the field may be headed.

Table 1.4 Summary of quantum software tools for quantum machine learning

Tool	Support device	Integration	Key features
PennyLane	IBM, Rigetti, IonQ, AQT, Xanadu	TensorFlow, PyTorch, JAX	Automatic differentiation for quantum circuits, extensive library of QML applications
Qiskit	IBM Quantum	Python, scikit-learn, PyTorch	Open-source, supports both simulation and real hardware execution
Cirq	Google Sycamore	TensorFlow Quantum, NumPy	Low-level quantum circuit design, fine control over gate execution
TFQ	Simulated backend, Cirq-compatible devices	TensorFlow	Integrated with TensorFlow for deep learning applications, quantum data processing
QuTiP	Simulated backends	NumPy, SciPy	Efficient simulations of noisy quantum systems, visualization tools for quantum dynamics

References

1. Feynman, R. P. (2017). Quantum mechanical computers. *Between Quantum and Cosmos, 16*(6), 523–548.
2. Watrous, J. (2008). Quantum computational complexity. arXiv preprint arXiv:0804.3401.
3. Shor, P. W. (1999). Polynomial-time algorithms for prime factorization and discrete logarithms on a quantum computer. *SIAM Review, 41*(2), 303–332.
4. Gidney, C., & Ekerå, M. (2021). How to factor 2048 bit RSA integers in 8 hours using 20 million noisy qubits. *Quantum, 5*, 433.
5. Biamonte, J., Wittek, P., Pancotti, N., Rebentrost, P., Wiebe, N., & Lloyd, S. (2017). Quantum machine learning. *Nature, 549*(7671), 195–202.
6. Kaplan, J., McCandlish, S., Henighan, T., Brown, T. B., Chess, B., Child, R., Gray, S., Radford, A., Wu, J., & Amodei, D. (2020). Scaling laws for neural language models. arXiv preprint arXiv:2001.08361
7. Benioff, P. (1982). Quantum mechanical hamiltonian models of turing machines. *Journal of Statistical Physics, 29*, 515–546.
8. Nielsen, M. A., & Chuang, I. L. (2011). *Quantum computation and quantum information: 10th anniversary edition* (2nd ed.). Cambridge University Press. ISBN 1-10-700217-6.
9. Raussendorf, R., & Briegel, H. J. (2001). A one-way quantum computer. *Physical Review Letters, 86*(22), 5188.
10. Albash, T., & Lidar, D. A. (2018). Adiabatic quantum computation. *Reviews of Modern Physics, 90*(1), 015002.
11. Kitaev, A. Y. (2003). Fault-tolerant quantum computation by anyons. *Annals of Physics, 303*(1), 2–30.
12. Grover, L. K. (1996). A fast quantum mechanical algorithm for database search. In *Proceedings of the Twenty-Eighth Annual ACM Symposium on Theory of Computing* (pp. 212–219).
13. Herman, D., Googin, C., Liu, X., Sun, Y., Galda, A., Safro, I., Pistoia, M., & Alexeev, Y. (2023). Quantum computing for finance. *Nature Reviews Physics, 5*(8), 450–465.

14. Santagati, R., Aspuru-Guzik, A., Babbush, R., Degroote, M., Gonzalez, L., Kyoseva, E., Moll, N., Oppel, M., Parrish, R. M., Rubin, N. C., et al. (2024). Drug design on quantum computers. *Nature Physics, 20*(4), 549–557.

15. Abbas, A., Ambainis, A., Augustino, B., Bärtschi, A., Buhrman, H., Coffrin, C., Cortiana, G., Dunjko, V., Egger, D. J., Elmegreen, B. G., et al. (2024). Challenges and opportunities in quantum optimization. *Nature Reviews Physics, 6*, 1–18.

16. Cross, A. W., Bishop, L. S., Sheldon, S., Nation, P. D., & Gambetta, J. M. (2019). Validating quantum computers using randomized model circuits. *Physical Review A, 100*(3), 032328.

17. Wack, A., Paik, H., Javadi-Abhari, A., Jurcevic, P., Faro, I., Gambetta, J. M., & Johnson, B. R. (2021). Quality, speed, and scale: Three key attributes to measure the performance of near-term quantum computers. arXiv preprint arXiv:2110.14108.

18. Kechedzhi, K., Isakov, S. V., Mandrà, S., Villalonga, B., Mi, X., Boixo, S., & Smelyanskiy, V. (2024). Effective quantum volume, fidelity and computational cost of noisy quantum processing experiments. *Future Generation Computer Systems, 153*, 431–441.

19. Arunachalam, S., & De Wolf, R. (2017). Guest column: A survey of quantum learning theory. *ACM Sigact News, 48*(2), 41–67.

20. Anshu, A., & Arunachalam, S. (2024). A survey on the complexity of learning quantum states. *Nature Reviews Physics, 6*(1), 59–69.

21. Harrow, A. W., & Montanaro, A. (2017). Quantum computational supremacy. *Nature, 549*(7671), 203–209.

22. Kim, Y., Eddins, A., Anand, S., Wei, K. X., Van Den Berg, E., Rosenblatt, S., Nayfeh, H., Wu, Y., Zaletel, M., Temme, K., et al. (2023). Evidence for the utility of quantum computing before fault tolerance. *Nature, 618*(7965), 500–505.

23. Schuld, M., Sinayskiy, I., & Petruccione, F. (2015). An introduction to quantum machine learning. *Contemporary Physics, 56*(2), 172–185.

24. Dunjko, V., & Briegel, H. J. (2018). Machine learning & artificial intelligence in the quantum domain: A review of recent progress. *Reports on Progress in Physics, 81*(7), 074001.

25. Carleo, G., Cirac, I., Cranmer, K., Daudet, L., Schuld, M., Tishby, N., Vogt-Maranto, L., & Zdeborová, L. (2019). Machine learning and the physical sciences. *Reviews of Modern Physics, 91*(4), 045002.

26. Cheng, B., Deng, X.-H., Gu, X., He, Y., Hu, G., Huang, P., Li, J., Lin, B.-C., Lu, D., Lu, Y., et al. (2023). Noisy intermediate-scale quantum computers. *Frontiers of Physics, 18*(2), 21308.

27. Huang, H.-L., Wu, D., Fan, D., & Zhu, X. (2020). Superconducting quantum computing: A review. *Science China Information Sciences, 63*, 1–32.

28. Bruzewicz, C. D., Chiaverini, J., McConnell, R., & Sage, J. M. (2019). Trapped-ion quantum computing: Progress and challenges. *Applied Physics Reviews, 6*(2), 021314.

29. Morgado, M., & Whitlock, S. (2021). Quantum simulation and computing with Rydberg-interacting qubits. *AVS Quantum Science, 3*(2), 1–36.

30. Preskill, J. (2018). Quantum computing in the NISQ era and beyond. *Quantum, 2*, 79.

31. Arute, F., Arya, K., Babbush, R., Bacon, D., Bardin, J. C., Barends, R., Biswas, R., Boixo, S., Brandao, F. G. S. L., Buell, D. A., et al. (2019). Quantum supremacy using a programmable superconducting processor. *Nature, 574*(7779), 505–510.

32. Wu, Y., Bao, W.-S., Cao, S., Chen, F., Chen, M.-C., Chen, X., Chung, T.-H., Deng, H., Du, Y., Fan, D., et al. (2021). Strong quantum computational advantage using a superconducting quantum processor. *Physical Review Letters, 127*(18), 180501.

33. Harrow, A. W., Hassidim, A., & Lloyd, S. (2009). Quantum algorithm for linear systems of equations. *Physical Review Letters, 103*(15), 150502.

34. Montanaro, A. (2016). Quantum algorithms: An overview. *NPJ Quantum Information, 2*(1), 1–8.

35. Dalzell, A. M., McArdle, S., Berta, M., Bienias, P., Chen, C.-F., Gilyén, A., Hann, C. T., Kastoryano, M. J., Khabiboulline, E. T., Kubica, A., et al. (2023). Quantum algorithms: A survey of applications and end-to-end complexities. arXiv preprint arXiv:2310.03011.

36. Lloyd, S., Mohseni, M., & Rebentrost, P. (September 2014). Quantum principal component analysis. *Nature Physics, 10*(9), 631–633. ISSN 1745-2481. https://doi.org/10.1038/nphys3029

37. Rebentrost, P., Mohseni, M., & Lloyd, S. (2014). Quantum support vector machine for big data classification. *Physical Review Letters, 113*(13), 130503.

38. Gilyén, A., Su, Y., Low, G. H., & Wiebe, N. (2019). Quantum singular value transformation and beyond: Exponential improvements for quantum matrix arithmetics. In *Proceedings of the 51st Annual ACM SIGACT Symposium on Theory of Computing*, STOC '19. ACM, June 2019. https://doi.org/10.1145/3313276.3316366

39. Chen, Z., Lu, Y., Wang, H., Liu, Y., & Li T. (2023). Quantum langevin dynamics for optimization. arXiv preprint arXiv:2311.15587.

40. Liu, J., Liu, M., Liu, J.-P., Ye, Z., Wang, Y., Alexeev, Y., Eisert, J., & Jiang, L. (2024). Towards provably efficient quantum algorithms for large-scale machine-learning models. *Nature Communications, 15*(1), 434.

41. Aaronson, S. (2015). Read the fine print. *Nature Physics, 11*(4), 291–293.

42. Giovannetti, V., Lloyd, S., & Maccone, L. (2008). Quantum random access memory. *Physical Review Letters, 100*(16), 160501.

43. Havlicek, V., Corcoles, A. D., Temme, K., Harrow, A. W., Kandala, A., Chow, J. M., & Gambetta, J. M. (2019). Supervised learning with quantum-enhanced feature spaces. *Nature, 567*(7747), 209–212.

44. Cerezo, M., Arrasmith, A., Babbush, R., Benjamin, S. C., Endo, S., Fujii, K., McClean, J. R., Mitarai, K., Yuan, X., Cincio, L., et al. (2021). Variational quantum algorithms. *Nature Reviews Physics, 3*(9), 625–644.

45. Li, W., & Deng, D.-L. (2022). Recent advances for quantum classifiers. *Science China Physics, Mechanics & Astronomy, 65*(2), 220301.

46. Tian, J., Sun, X., Du, Y., Zhao, S., Liu, Q., Zhang, K., Yi, W., Huang, W., Wang, C., Wu, X., et al. (2023). Recent advances for quantum neural networks in generative learning. *IEEE Transactions on Pattern Analysis and Machine Intelligence, 45*(10), 12321–12340

47. Bharti, K., Cervera-Lierta, A., Kyaw, T. H., Haug, T., Alperin-Lea, S., Anand, A., Degroote, M., Heimonen, H., Kottmann, J. S., enke, T., et al. (2022). Noisy intermediate-scale quantum algorithms. *Reviews of Modern Physics, 94*(1), 015004.

48. Cerezo, M., Verdon, G., Huang, H.-Y., Cincio, L., & Coles, P. J. (2022). Challenges and opportunities in quantum machine learning. *Nature Computational Science, 2*(9), 567–576.

49. Du, Y., Huang, T., You, S., Hsieh, M.-H., & Tao, D. (2022). Quantum circuit architecture search for variational quantum algorithms. *NPJ Quantum Information, 8*(1), 62.

50. Stokes, J., Izaac, J., Killoran, N., & Carleo, G. (2020). Quantum natural gradient. *Quantum, 4*, 269.

51. Sim, S., Romero, J., Gonthier, J. F., & Kunitsa, A. A. (2021). Adaptive pruning-based optimization of parameterized quantum circuits. *Quantum Science and Technology, 6*(2), 025019.

52. Wang, X., Liu, J., Liu, T., Luo, Y., Du, Y., & Tao, D. (2023). Symmetric pruning in quantum neural networks. In *The Eleventh International Conference on Learning Representations.* https://openreview.net/forum?id=K96AogLDT2K

53. Lu, S., Duan, L.-M., & Deng, D.-L. (Aug 2020). Quantum adversarial machine learning. *Physical Review Research, 2*, 033212. https://doi.org/10.1103/PhysRevResearch.2.033212

54. Jiang, W., Lu, Z., & Deng, D.-L. (May 2022). Quantum continual learning overcoming catastrophic forgetting. *Chinese Physics Letters, 39*(5), 050303. https://doi.org/10.1088/0256-307X/39/5/050303

55. Zhang, C., Lu, Z., Zhao, L., Xu, S., Li, W., Wang, K., Chen, J., Wu, Y., Jin, F., Zhu, X., et al. (2024). Quantum continual learning on a programmable superconducting processor. arXiv preprint arXiv:2409.09729.

56. Du, Y., Hsieh, M.-H., Liu, T., Tao, D., & Liu, N. (2021). Quantum noise protects quantum classifiers against adversaries. *Physical Review Research, 3*(2), 023153.

57. Watkins, W. M., Chen, S. Y.-C., & Yoo, S. (2023). Quantum machine learning with differential privacy. *Scientific Reports, 13*(1), 2453. https://doi.org/10.1038/s41598-022-24082-z

58. Du, Y., Qian, Y., Wu, X., & Tao, D. (2022). A distributed learning scheme for variational quantum algorithms. *IEEE Transactions on Quantum Engineering, 3*, 1–16.

59. Sheng, Y.-B., & Zhou, L. (2017). Distributed secure quantum machine learning. *Science Bulletin, 62*(14), 1025–1029.
60. Ren, C., Yan, R., Zhu, H., Yu, H., Xu, M., Shen, Y., Xu, Y., Xiao, M., Dong, Z. Y., Skoglund, M., et al. (2023). Towards quantum federated learning. arXiv preprint arXiv:2306.09912.
61. Pira, L., & Ferrie, C. (2024). On the interpretability of quantum neural networks. *Quantum Machine Intelligence, 6*(2), 52.
62. Banchi, L., Pereira, J. L., Jose, S. T., & Simeone, O. (2023). Statistical complexity of quantum learning. *Advanced Quantum Technologies*, 2300311.
63. Bergholm, V., Izaac, J., Schuld, M., Gogolin, C., Ahmed, S., Ajith, V., Alam, M. S., Alonso-Linaje, G., AkashNarayanan, B., Asadi, A., et al. (2018). Pennylane: Automatic differentiation of hybrid quantum-classical computations. arXiv preprint arXiv:1811.04968.
64. Javadi-Abhari, A., Treinish, M., Krsulich, K., Wood, C. J., Lishman, J., Gacon, J., Martiel, S., Nation, P. D., Bishop, L. S., Cross, A. W., et al. (2024). Quantum computing with qiskit. arXiv preprint arXiv:2405.08810.
65. Cirq Developers. (May 2024). Cirq. https://doi.org/10.5281/zenodo.11398048.
66. Broughton, M., Verdon, G., McCourt, T., Martinez, A. J., Yoo, J. H., Isakov, S. V., Massey, P., Halavati, R., Niu, M. Y., Zlokapa, A., et al. (2020). Tensorflow quantum: A software framework for quantum machine learning. arXiv preprint arXiv:2003.02989.

Chapter 2
Basics of Quantum Computing

Abstract This chapter introduces the fundamental concepts of quantum computation, such as quantum states, quantum circuits, and quantum measurements, along with key topics in quantum machine learning, including quantum read-in, quantum read-out, and quantum linear algebra. These foundational elements are essential for understanding quantum machine learning algorithms and will be repeatedly referenced throughout the subsequent chapters. This chapter is organized into six sections: Sect. 2.1 introduces quantum bits and their mathematical representations; Sect. 2.2 covers quantum circuits, including quantum gates, quantum channels, and quantum measurements; Sect. 2.3 discusses how to encode classical data into quantum systems and extract classical information from quantum states; Sect. 2.4 explores concepts in quantum linear algebra; Sect. 2.5 provides practical coding exercises to reinforce these concepts; and finally, Sect. 2.6 presents recent advancements in efficient quantum read-in and read-out techniques, as well as developments in quantum linear algebra for further exploration.

This chapter introduces the fundamental concepts of quantum computation, including quantum states, quantum circuits, and quantum measurements. Key topics in quantum machine learning are also presented, such as quantum read-in methods, quantum read-out methods, and quantum linear algebra. These foundational elements are essential for understanding quantum machine learning algorithms and are referenced throughout the subsequent chapters.

This chapter is organized as follows. Section 2.1 introduces quantum bits and their mathematical representations. Section 2.2 covers quantum circuits, including quantum gates, quantum channels, and quantum measurements. Section 2.3 discusses how to encode classical data into quantum systems and extract classical information from quantum states. Section 2.4 delves into quantum linear algebra; Sect. 2.5 provides practical coding exercises to reinforce these concepts. Finally, Sect. 2.6 presents recent advancements in efficient quantum read-in and read-out techniques for further exploration.

© The Author(s), under exclusive license to Springer Nature Singapore Pte Ltd. 2025   23
Y. Du et al., *A Gentle Introduction to Quantum Machine Learning*,
https://doi.org/10.1007/978-981-95-1284-3_2

2.1 From Classical Bits to Quantum Bits

In this section, quantum bits (qubits) are defined and the mathematical tools used to describe quantum states are presented. The discussion begins with classical bits and then transitions to their quantum counterparts. Interested readers are recommended to consult the textbook [1] for the detailed explanations.

2.1.1 Classical Bits

In classical computing, a bit is the basic unit of information, which can exist in one of two distinct states: 0 or 1. Each bit holds a definite value at any given time. When multiple classical bits are used together, they can represent more complex information. For instance, a set of three bits can represent $2^3 = 8$ distinct states, ranging from 000 to 111.

2.1.2 Quantum Bits (Qubits)

Analogous to the role of "bit" in classical computation, the basic element in quantum computation is the quantum bit (*qubit*). The representation of single-qubit states is introduced first, followed by an extension to two-qubit and multi-qubit states.

Single-Qubit State A single-qubit state can be represented by a two-dimensional vector with unit length. Mathematically, a qubit state can be written as

$$a = \begin{bmatrix} a_1 \\ a_2 \end{bmatrix} \in \mathbb{C}^2 , \tag{2.1}$$

where $|a_1|^2 + |a_2|^2 = 1$ satisfies the *normalization constraint*. Following conventions in quantum theory, Dirac notation is used to represent vectors [1]. That is, the vector a is denoted by $|a\rangle$ (named "*ket*") with

$$|a\rangle = a_1|0\rangle + a_2|1\rangle , \tag{2.2}$$

where $|0\rangle \equiv e_0 \equiv \begin{bmatrix} 1 \\ 0 \end{bmatrix}$ and $|1\rangle \equiv e_1 \equiv \begin{bmatrix} 0 \\ 1 \end{bmatrix}$ are two computational (unit) basis states. In this representation, the coefficients a_1 and a_2 are referred to as *amplitudes*. The probabilities of obtaining the outcomes 0 or 1 upon measurement of the qubit are given by $|a_1|^2$ and $|a_2|^2$, respectively. The normalization constraint ensures that these probabilities always sum to one, as required by the probabilistic nature of quantum mechanics. In addition, the conjugated transpose of a, i.e., $a^\dagger$, is denoted by $\langle a|$ (named "*bra*") with

$$\langle a| = a_1^* \langle 0| + a_2^* \langle 1| \in \mathbb{C}^2 , \tag{2.3}$$

where $\langle 0| \equiv e_0^\top \equiv [1, 0]$, $\langle 1| \equiv e_1^\top \equiv [0, 1]$, and the symbol "$\top$" denotes the transpose operation.

The physical interpretation of coefficients $\{a_i\}$ is *probability amplitudes*. Namely, when information needs to be extracted from the qubit state $|a\rangle$ into the classical form, quantum measurements are applied to this state, where the probability of sampling the basis $|0\rangle$ ($|1\rangle$) is $|a_1|^2$ ($|a_2|^2$). Recall that the classical bit only permits the deterministic status with "0" or "1," while the qubit state in Eq. (2.2) is the *superposition* of the two status "$|0\rangle$" and "$|1\rangle$."

> **Remark**
>
> The *quantum superposition* leads to a distinct power between quantum and classical computation, where the former can accomplish certain tasks with provable advantages.

Two-Qubit State   The two qubits obey the tensor product rule, i.e.,

$$\begin{bmatrix} x_1 \\ x_2 \end{bmatrix} \otimes \begin{bmatrix} y_1 \\ y_2 \end{bmatrix} = \begin{bmatrix} x_1 \begin{bmatrix} y_1 \\ y_2 \end{bmatrix} \\ x_2 \begin{bmatrix} y_1 \\ y_2 \end{bmatrix} \end{bmatrix} = \begin{bmatrix} x_1 y_1 \\ x_1 y_2 \\ x_2 y_1 \\ y_2 y_2 \end{bmatrix} , \tag{2.4}$$

which differs from the classical bits yielding the Cartesian product rule.

For instance, let the first qubit follow Eq. (2.2) and the second qubit state be $|b\rangle = b_1|0\rangle + b_2|1\rangle$ with $|b_1|^2 + |b_2|^2 = 1$. The two-qubit state formed by $|a\rangle$ and $|b\rangle$ is defined as

$$|a\rangle \otimes |b\rangle = a_1 b_1 |0\rangle \otimes |0\rangle + a_1 b_2 |0\rangle \otimes |1\rangle + a_2 b_1 |1\rangle \otimes |0\rangle + a_2 b_2 |1\rangle \otimes |1\rangle \in \mathbb{C}^4 , \tag{2.5}$$

where the computational basis follows $|0\rangle \otimes |0\rangle \equiv \begin{bmatrix} 1 \\ 0 \\ 0 \\ 0 \end{bmatrix}$, $|0\rangle \otimes |1\rangle \equiv \begin{bmatrix} 0 \\ 1 \\ 0 \\ 0 \end{bmatrix}$, $|1\rangle \otimes |0\rangle \equiv \begin{bmatrix} 0 \\ 0 \\ 1 \\ 0 \end{bmatrix}$, $|1\rangle \otimes |1\rangle \equiv \begin{bmatrix} 0 \\ 0 \\ 0 \\ 1 \end{bmatrix}$, and the coefficients satisfy $\sum_{i=1}^{2} \sum_{j=1}^{2} |a_i b_j|^2 = 1$.

Remark
For ease of notations, the state $|a\rangle \otimes |b\rangle$ can be simplified as $|ab\rangle$, $|a, b\rangle$, or $|a\rangle|b\rangle$. These notations will be used *interchangeably* throughout the book.

Example 2.1 A typical example of a two-qubit state is the *Bell state*, which represents a maximally entangled quantum state of two qubits. There are four types of Bell states, expressed ass

$$|\phi^+\rangle = \frac{1}{\sqrt{2}} \left(|00\rangle + |11\rangle\right),$$

$$|\phi^-\rangle = \frac{1}{\sqrt{2}} \left(|00\rangle - |11\rangle\right),$$

$$|\psi^+\rangle = \frac{1}{\sqrt{2}} \left(|01\rangle + |10\rangle\right),$$

$$|\psi^-\rangle = \frac{1}{\sqrt{2}} \left(|01\rangle - |10\rangle\right). \tag{2.6}$$

Each Bell state is a superposition of two computational basis states in the four-dimensional Hilbert space.

Multi-qubit State The above two-qubit case can now be generalized to the N-qubit case with $N > 2$. In particular, an N-qubit state $|\psi\rangle$ is a 2^N-dimensional vector with

$$|\psi\rangle = \sum_{i=1}^{2^N} c_i |i\rangle \in \mathbb{C}^{2^N}, \tag{2.7}$$

where the coefficients satisfy the normalization constraint $\sum_{i=1}^{2^N} |c_i|^2 = 1$ and the symbol "i" of the computational basis $|i\rangle$ refers to a bit-string with $i \in \{0, 1\}^N$. As with the single-qubit case, the physical interpretation of coefficients $\{c_i\}$ is probability amplitudes, where the probability to sample the bit-string "i" is $|c_i|^2$. When the number of nonzero entries in $c = [c_1, \ldots, c_i, \ldots, c_{2^N}]^\top$ is larger than one, which implies that different bit-strings are coexisting coherently, the state $|\psi\rangle$ is called *in superposition*.

Remark

In quantum computing, a basis state $|i\rangle$ refers to a computational basis state in the Hilbert space of a quantum system. For an N-qubit system, the computational basis states are represented as $|i\rangle \in \{|0\cdots 0\rangle, |0\cdots 1\rangle, \ldots, |1\cdots 1\rangle\}$, where i is the binary representation of the state index. These states form an orthonormal basis of the 2^N-dimensional Hilbert space, satisfying

$$\langle i|j\rangle = \delta_{ij}, \forall i, j \in \left[2^N\right]. \tag{2.8}$$

These basis states are fundamental for representing and analyzing quantum states, as any arbitrary quantum state can be expressed as a linear combination of these basis states.

Moreover, the size of c exponentially scales with the number of qubits N, attributed to the tensor product rule. This exponential dependence is an indispensable factor to achieve quantum supremacy [2], since it is extremely expensive and even intractable to record all information of c by classical devices for the modest number of qubits, e.g., $N > 100$.

Entangled Multi-qubit State A fundamental phenomenon in multi-qubit quantum systems is *entanglement*, which represents a nonclassical correlation between quantum systems that cannot be explained by classical physics. As proved by Jozsa and Linden [3], quantum entanglement is an indispensable component to offer an exponential speedup over classical computation. A representative example is Shor's algorithm, which utilizes entanglement to attain an exponential speedup over any classical factoring algorithm. In an entangled quantum state, the state of one qubit cannot be fully described independently of the other qubits, even if they are spatially separated. The formal definition of entanglement for states in Dirac notation is as follows.

Definition 2.1 (Entanglement for States in Dirac Notation) An N-qubit state $|\psi\rangle \in \mathbb{C}^{2^N}$ is *entangled* if it cannot be expressed as the tensor product of states of its subsystems A and B:

$$|\psi\rangle \neq |\psi_a\rangle \otimes |\psi_b\rangle, \quad \forall |\psi_a\rangle \in \mathbb{C}^{2^{N_A}}, |\psi_b\rangle \in \mathbb{C}^{2^{N_B}}, N_A + N_B = N. \tag{2.9}$$

If the state can be expressed in this form, it is referred to as *seperable*.

Example 2.2 (GHZ State) A typical example of an entangled N-qubit state is the Greenberger–Horne–Zeilinger (GHZ) state [4]. GHZ state is a generalization of the two-qubit Bell state (see Example 2.1) to a maximally entangled N-qubit state. The general form of an N-qubit GHZ state is

$$|\text{GHZ}_N\rangle = \frac{1}{\sqrt{2}} \left(|0\rangle^{\otimes N} + |1\rangle^{\otimes N} \right).$$
(2.10)

For $N = 3$, the GHZ state is

$$|\text{GHZ}_3\rangle = \frac{1}{\sqrt{2}} \left(|000\rangle + |111\rangle \right).$$
(2.11)

A key property of the entangled states (e.g., Bell states and GHZ states) is that measuring one qubit determines the outcome of measuring the other qubit, reflecting their strong quantum correlation.

2.1.3 Density Matrix

Another description of quantum states is through *density matrix* or *density operators*. The reason for establishing density operators instead of Dirac notations arises from the imperfection of physical systems. Specifically, Dirac notations introduced in Sect. 2.1.2 are used to describe "ideal" quantum states (i.e., *pure states*), where the operated qubits are isolated from the environment. Alternatively, when the operated qubits interact with the environment unavoidably, the density operators are employed to describe the behavior of quantum states living in this open system. As such, density operators describe more general quantum states.

Mathematically, an N-qubit density operator, denoted by $\rho \in \mathbb{C}^{2^N \times 2^N}$, presents a mixture of m quantum pure states $|\psi_i\rangle \in \mathbb{C}^{2^N}$ with probability $p_i \in [0, 1]$ and $\sum_{i=1}^{m} p_i = 1$. That is,

$$\rho = \sum_{i=1}^{m} p_i \rho_i ,$$
(2.12)

where $\rho_i = |\psi_i\rangle \langle\psi_i| \in \mathbb{C}^{2^N \times 2^N}$ is the outer product of the pure state $|\psi_i\rangle$. The outer product of two vectors $|u\rangle, |v\rangle \in \mathbb{C}^n$ is expressed as

$$|u\rangle \langle v| = \begin{bmatrix} u_1 \\ u_2 \\ \vdots \\ u_n \end{bmatrix} \begin{bmatrix} v_1^* & v_2^* & \cdots & v_n^* \end{bmatrix} = \begin{bmatrix} u_1 v_1^* & u_1 v_2^* & \cdots & u_1 v_n^* \\ u_2 v_1^* & u_2 v_2^* & \cdots & u_2 v_n^* \\ \vdots & \vdots & \ddots & \vdots \\ u_n v_1^* & u_n v_2^* & \cdots & u_n v_n^* \end{bmatrix}, \tag{2.13}$$

where u_i and v_i^* are the element of $|u\rangle$ and the conjugate transpose $\langle v|$, respectively.

From the perspective of computer science, the density operator ρ is just a *positive semi-definite matrix* with trace-preserving, i.e., $0 \preceq \rho$ and $\mathrm{Tr}(\rho) = 1$.

Definition 2.2 (Positive Semi-definite Matrix) A matrix $A \in \mathbb{C}^{n \times n}$ is positive semi-definite (PSD) if it satisfies the following conditions:

1. A is Hermitian: $A = A^\dagger$.
2. For any nonzero vector $|v\rangle \in \mathbb{C}^n$, $\langle v| A |v\rangle \geq 0$, where $\langle v| A |v\rangle$ represents the quadratic form of A with respect to $|v\rangle$.

When $m = 1$, the density operator ρ amounts to a pure state with $\rho = |\psi_1\rangle \langle\psi_1|$. When $m > 1$, the density operator ρ describes a "mixed" quantum state, where the rank of ρ is larger than 1. A simple criterion distinguishes pure states from mixed states: a pure state $m = 1$ yields $\mathrm{Tr}(\rho^n) = \mathrm{Tr}(\rho) = 1$ for any $n \in \mathbb{N}_+$. Conversely, a mixed state with $m > 1$ satisfies $\mathrm{Tr}(\rho^n) < \mathrm{Tr}(\rho) = 1$ for any $n \in \mathbb{N}_+ \setminus \{1\}$, as its rank is greater than 1. Similar to Definition 2.1 for entanglement of pure states, the entanglement of mixed states can also be defined.

Definition 2.3 (Entanglement for Mixed States) Let ρ be a density operator acting on a composite Hilbert space $\mathcal{H}_A \otimes \mathcal{H}_B$. The state ρ is said to be *entangled* if it cannot be expressed as

$$\rho = \sum_i p_i \, \rho_A^{(i)} \otimes \rho_B^{(i)}, \tag{2.14}$$

where $p_i \geq 0$, $\sum_i p_i = 1$, and $\rho_A^{(i)}$ and $\rho_B^{(i)}$ are density operators on $\mathcal{H}_A$ and $\mathcal{H}_B$, respectively. If ρ can be written in this form, it is called *separable*.

Example 2.3 (Density Matrix Representations)

(i) Consider the single-qubit pure state $|\psi\rangle = \frac{1}{\sqrt{2}}(|0\rangle + |1\rangle)$. The corresponding density operator is

$$\rho = |\psi\rangle \langle\psi| = \frac{1}{2} \begin{bmatrix} 1 & 1 \\ 1 & 1 \end{bmatrix}.$$

(continued)

Example 2.3 (continued)

Here, $\mathrm{Tr}\left(\rho^2\right) = \mathrm{Tr}(\rho) = 1$, confirming that it is a pure state.

(ii) Consider the classical probabilistic mixture of $|0\rangle$ and $|1\rangle$, each with equal probability $p = 0.5$. The density operator is

$$\rho = 0.5|0\rangle \langle 0| + 0.5|1\rangle \langle 1| = \frac{1}{2}\begin{bmatrix} 1 & 0 \\ 0 & 1 \end{bmatrix}.$$

In this case, $\mathrm{Tr}\left(\rho^2\right) = 0.5 < \mathrm{Tr}(\rho) = 1$, indicating it is a mixed state.

2.2 From Digital Logic Circuit to Quantum Circuit Model

To process quantum states, quantum computation is introduced, with the quantum circuit model serving as a fundamental framework. This section begins with classical computation in Sect. 2.2.1 and transitions to details about the quantum circuit model in Sect. 2.2.2, including quantum gates, quantum channel, and quantum measurements.

2.2.1 Classical Digital Logic Circuit

Digital logic circuits are the foundational building blocks of classical computing systems. They process classical bits by performing logic operations through logic gates. In this subsection, the essential components of digital logic circuits and their functionality are introduced, followed by a discussion of how these classical circuits relate to quantum circuits.

2.2.1.1 Logic Gates

Logic gates are the basic components of a digital circuit. They take binary inputs, represented as 0 or 1, and produce a binary output based on a predefined logic operation. The most common logic gates are summarized below.

1. **NOT Gate**: This gate inverts the input bit, i.e., it produces 1 if the input is 0, and vice versa. Its truth table is shown in Table 2.1.
2. **AND Gate**: Produces an output of 1 only if both input bits are 1; otherwise, it outputs 0. The truth table is shown in Table 2.2.

Table 2.1 Input-output mapping of the NOT gate

Input (A)	Output (NOT A)
0	1
1	0

Table 2.2 Input-output mapping of the AND gate

Input (A)	Input (B)	Output (A AND B)
0	0	0
0	1	0
1	0	0
1	1	1

Table 2.3 Input-output mapping of the OR gate

Input (A)	Input (B)	Output (A OR B)
0	0	0
0	1	1
1	0	1
1	1	1

Table 2.4 Input-output mapping of the XOR gate

Input (A)	Input (B)	Output (A XOR B)
0	0	0
0	1	1
1	0	1
1	1	0

3. **OR Gate**: Outputs 1 if at least one input is 1. The truth table is shown in Table 2.3.
4. **XOR Gate**: Produces an output of 1 if the inputs are different and 0 otherwise. The truth table is shown in Table 2.4.

These logic gates can be combined in various configurations to build more complex circuits capable of performing arbitrary arithmetic operations.

2.2.1.2   Circuit Design and Universality

A classical digital logic circuit is composed of interconnected gates designed to perform specific tasks, such as addition or multiplication. A key property of these circuits is **universality**, meaning any logic function can be implemented using a finite set of gates. For example, the **NAND Gate** (NOT AND) and **NOR Gate** (NOT OR) are universal gates. Any other logic operation can be constructed using only NAND or NOR gates [5].

2.2.2 Quantum Circuit

Classical digital logic circuits provide the essential framework for understanding computation. While classical circuits operate on bits and perform deterministic operations, quantum circuits manipulate qubits and involve probabilistic behavior. The concepts of logic gates, circuit design, and universality lay the groundwork for transitioning to quantum circuits introduced in this subsection.

2.2.2.1 Quantum Gate

Recall that the computational toolkit for classical computers is logic gates, e.g., NOT, AND, OR, and XOR, which are applied to the single bit or multiple bits to accomplish computation. Similarly, the computational toolkit for quantum computers (or quantum circuits) is **quantum gate**, which *operates on qubits* introduced in Sect. 2.1.2 to complete the computation. Both single-qubit and multi-qubit gates are introduced in the following.

Single-Qubit Gates Single-qubit gates control the evolution of the single-qubit state $|a\rangle$. Due to the law of quantum mechanics, the evolved state should satisfy the normalization constraint. The implication of this constraint is that the evolution must be a unitary operation. Concretely, denoted $U \in \mathbb{C}^{2\times2}$ as a linear operator and the evolved state as

$$|\hat{a}\rangle := U|a\rangle = \hat{a}_1|0\rangle + \hat{a}_2|1\rangle \in \mathbb{C}^2 , \tag{2.15}$$

the summation of coefficients $|\hat{a}_1|^2 + |\hat{a}_2|^2 = \langle\hat{a}|\hat{a}\rangle = \langle a|U^\dagger U|a\rangle$ is equal to 1 if and only if U is unitary with $U^\dagger U = U U^\dagger = \mathbb{I}_2$. The symbol "$\dagger$" denotes the conjugate transpose operation. Under the density operator representation, the evolution of $|a\rangle$ yields

$$\hat{\rho} = U\rho U^\dagger , \tag{2.16}$$

where $\hat{\rho} = |\hat{a}\rangle\langle\hat{a}|$ and $\rho = |a\rangle\langle a|$.

Several common single-qubit gates, including Pauli-X, Pauli-Y, Pauli-Z, Hadamard, and rotational single-qubit gates about the X, Y, and Z axes (RX, RY, RZ), are illustrated in Fig. 2.1. According to Theorem 4.1 in [1], any unitary operation on a single qubit can be decomposed into a sequence of rotations as

$$U = \mathrm{RZ}(\alpha)\,\mathrm{RY}(\beta)\,\mathrm{RZ}(\gamma), \tag{2.17}$$

where $\alpha, \beta, \gamma \in [0, 2\pi)$, up to a global phase shift.

Quantum gate	Matrix form	Circuit representation
Pauli-X (X)	$\begin{pmatrix} 0 & 1 \\ 1 & 0 \end{pmatrix}$	$-\boxed{X}-$
Pauli-Y (Y)	$\begin{pmatrix} 0 & -i \\ i & 0 \end{pmatrix}$	$-\boxed{Y}-$
Pauli-Z (Z)	$\begin{pmatrix} 1 & 0 \\ 0 & -1 \end{pmatrix}$	$-\boxed{Z}-$
Hadamard (H)	$\dfrac{1}{\sqrt{2}}\begin{pmatrix} 1 & 1 \\ 1 & -1 \end{pmatrix}$	$-\boxed{H}-$
Controlled-Z (CZ)	$\begin{pmatrix} 1 & 0 & 0 & 0 \\ 0 & 1 & 0 & 0 \\ 0 & 0 & 1 & 0 \\ 0 & 0 & 0 & -1 \end{pmatrix}$	$\boxed{Z}$
RX(θ)	$\begin{pmatrix} \cos\left(\frac{\theta}{2}\right) & -i\sin\left(\frac{\theta}{2}\right) \\ -i\sin\left(\frac{\theta}{2}\right) & \cos\left(\frac{\theta}{2}\right) \end{pmatrix}$	$-\boxed{R_x(\theta)}-$
RY(θ)	$\begin{pmatrix} \cos\left(\frac{\theta}{2}\right) & -\sin\left(\frac{\theta}{2}\right) \\ \sin\left(\frac{\theta}{2}\right) & \cos\left(\frac{\theta}{2}\right) \end{pmatrix}$	$-\boxed{R_y(\theta)}-$
RZ(θ)	$\begin{pmatrix} e^{-i\theta/2} & 0 \\ 0 & e^{i\theta/2} \end{pmatrix}$	$-\boxed{R_z(\theta)}-$
SWAP	$\begin{pmatrix} 1 & 0 & 0 & 0 \\ 0 & 0 & 1 & 0 \\ 0 & 1 & 0 & 0 \\ 0 & 0 & 0 & 1 \end{pmatrix}$	$\times$
Controlled-NOT (CNOT)	$\begin{pmatrix} 1 & 0 & 0 & 0 \\ 0 & 1 & 0 & 0 \\ 0 & 0 & 0 & 1 \\ 0 & 0 & 1 & 0 \end{pmatrix}$	$\oplus$

Fig. 2.1 The summarization of quantum gates

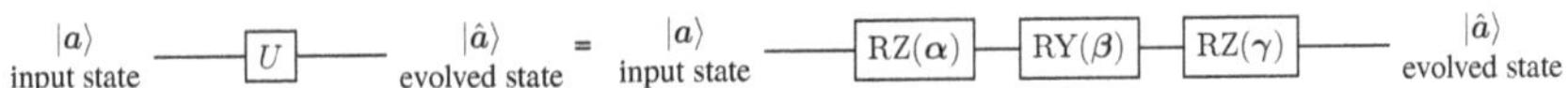

Fig. 2.2 The evolution of the single-qubit state decomposed into the quantum gates

The evolution from $|a\rangle$ to $|\hat{a}\rangle$ can be visualized using a quantum circuit diagram, as illustrated in Fig. 2.2. The wire in the circuit represents a qubit, which evolves from the initial state $|a\rangle$ on the left to the final state $|\hat{a}\rangle$ on the right. Gates are applied sequentially from left to right along the wire.

Remark

The circuit model serves as a foundational framework for describing quantum computation due to its *intuitive* and *modular* nature, making it accessible for researchers and practitioners transitioning from classical to quantum computing. First, the circuit model provides a standardized graphical language to represent complex quantum algorithms, enabling clear visualization of the computational flow and interactions among qubits. Second, the modularity of the circuit model allows quantum operations to be easily decomposed into a predefined gate set, ensuring compatibility across different quantum hardware architectures.

Multi-qubit Gates The evolution of the N-qubit quantum state can be effectively generalized by the single-qubit case. That is, the unitary operator $U \in \mathbb{C}^{2^N \times 2^N}$ evolves an N-qubit state $|\psi\rangle$ in Eq. (2.7) as

$$|\widehat{\psi}\rangle = U|\psi\rangle \in \mathbb{C}^{2^N} . \tag{2.18}$$

The evolution of $|\psi\rangle$ under the density operator representation is denoted by $\hat{\rho} = U\rho U^{\dagger}$, where $\hat{\rho} = |\widehat{\psi}\rangle \langle \widehat{\psi}|$ and $\rho = |\psi\rangle \langle \psi|$.

Remark

In the view of computer science, the quantum (logic) gates in Fig. 2.1 are well-designed matrices with the following properties. First, all quantum gates are unitary (e.g., $XX^{\dagger} = \mathbb{I}_2$). Second, X, Y, Z, H gates have the fixed form with size 2×2; CNOT, CZ, and SWAP gates have the fixed form with size 4×4. Third, $RX(\theta)$, $RY(\theta)$, $RZ(\theta)$ gates are matrices controlled by a single variable θ.

Figure 2.1 includes two significant multi-qubit gates: the controlled-Z (CZ) gate and the controlled-NOT (CNOT) gate. For instance, the CNOT gate operates on two qubits: a *control* qubit (top line) and a *target* qubit (bottom line). If the control qubit is 0, the target qubit remains unchanged; if the control qubit is 1, the target qubit is flipped.

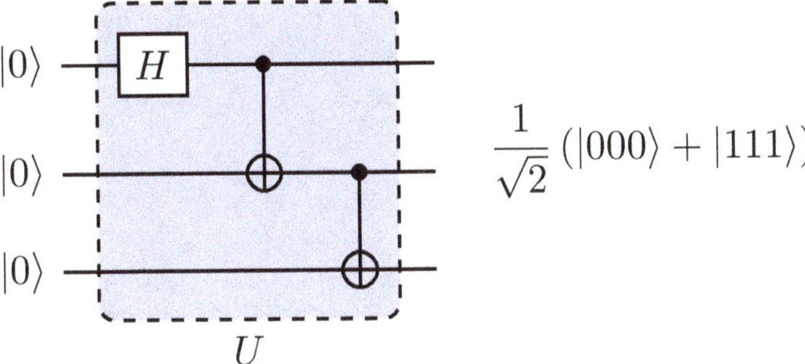

Fig. 2.3 The decomposition of the multi-qubit circuit U in the case of $N = 3$

Example 2.4 (State Evolved by Multi-qubit Gates) Figure 2.3 illustrates the evolution of a 3-qubit state $|\psi\rangle$ under a multi-qubit circuit consisting of multi-qubit gates. Each wire represents a qubit, and the evolution occurs from left to right. Starting with the initial state $|\psi\rangle = |000\rangle$, a Hadamard gate is applied to the first qubit, followed by two CNOT gates. That is, the first one acts on the first and second qubits, and the other acts on the second and third qubits. The final evolved state, shown on the right, is the GHZ state introduced in Example 2.2, i.e., $|\widehat{\psi}\rangle = U|\psi\rangle = \frac{1}{\sqrt{2}}(|000\rangle + |111\rangle)$. The entire unitary operation can be represented as

$$U = (\mathbb{I}_2 \otimes \text{CNOT})(\text{CNOT} \otimes \mathbb{I}_2)(\text{H} \otimes \mathbb{I}_4). \tag{2.19}$$

Remark
The CNOT gate plays a pivotal role in quantum computing due to its unique ability to generate entangled states, such as the Bell states and GHZ states presented in Examples 2.1 and 2.2. Besides, the CNOT gate is one of the most commonly implemented gates on quantum hardware. Its design and optimization directly impact the fidelity and scalability of quantum systems.

A Universal Quantum Gate Set While many single and multi-qubit gates exist, it is sufficient to use a *universal set of gates* to construct any unitary operation. As proved in Chapter 4.5.2 of Ref. [1], any unitary operator U in Eq. (2.18) can be decomposed into the single-qubit and two-qubit gates with a certain arrangement.

Fact 2.1 (Solovay-Kitaev Theorem, [6]) *Consider a fixed universal gate set $\mathcal{G}$, which generates a dense group* $\mathrm{SU}(d)$. *Then, any unitary operator* $U \in \mathrm{SU}(d)$ *can be approximated to an arbitrary precision* $\epsilon > 0$ *by a finite sequence of gates from* $\mathcal{G}$. *Formally, there exists a decomposition such that*

$$\left\| U - \prod_{l=1}^{L} G_l \right\|_{op} \leq \epsilon, \quad G_l \in \mathcal{G}, \quad L \in \mathbb{N}, \tag{2.20}$$

where $\|\cdot\|_{op}$ *is the operator norm which is the largest singular value of a matrix and* L *is the required number of gates that scales as*

$$L = O\left(\log^c(1/\epsilon)\right), \tag{2.21}$$

with $c \approx 4$.

A commonly used universal gate set includes single-qubit rotations $\mathrm{RX}(\theta)$, $\mathrm{RY}(\theta)$, $\mathrm{RZ}(\theta)$, and two-qubit gates such as the CNOT gate. As illustrated in Fig. 1.1, any ideal quantum computation can be represented by a unitary operator. This universal gate set provides a practical and foundational toolkit for implementing arbitrary quantum algorithms.

2.2.2.2 Quantum Channels

Analogous to the unitary operation describing the evolution of quantum states in the closed system, the quantum channel formalizes the evolution of quantum states in the open system. Refer to the textbook [7] for more details.

Mathematically, every quantum channel $\mathcal{N}(\cdot)$ can be treated as a linear, completely positive, and trace-preserving map (CPTP map).

Definition 2.4 (CPTP Map) Denote $\mathcal{L}(\mathcal{H})$ as the space of square linear operators acting on the Hilbert space $\mathcal{H}$. Then, $\mathcal{N}(\cdot)$ is a CPTP map if the following conditions are satisfied:

- Linearity means that for any $X_A, Y_A \in \mathcal{L}(\mathcal{H}_A)$ and $a, b \in \mathbb{C}$, $\mathcal{N}(aX_A + bY_A) = a\mathcal{N}(X_A) + b\mathcal{N}(Y_A)$.
- A linear map $\mathcal{N} : \mathcal{L}(\mathcal{H}_A) \to \mathcal{L}(\mathcal{H}_B)$ is a positive map if $\mathcal{N}(X_A)$ is positive semi-definite for all positive semi-definite operators $X_A \in \mathcal{L}(\mathcal{H}_A)$. Furthermore, a linear map $\mathcal{N} : \mathcal{L}(\mathcal{H}_A) \to \mathcal{L}(\mathcal{H}_B)$ is completely positive if $\mathbb{I}_R \otimes \mathcal{N}$ is a positive map for any size of R.
- Trace preservation means that $\mathrm{Tr}(\mathcal{N}(X_A)) = \mathrm{Tr}(X_A)$ for any $X_A \in \mathcal{L}(\mathcal{H}_A)$.

A quantum channel can be represented by the Choi–Kraus decomposition [1]. Mathematically, let $\mathcal{L}(\mathcal{H}_A, \mathcal{H}_B)$ denote the space of linear operators taking $\mathcal{H}_A$ to

$\mathcal{H}_B$. The Choi–Kraus decomposition of the quantum channel $\mathcal{N}(\cdot) : \mathcal{L}(\mathcal{H}_A) \to \mathcal{L}(\mathcal{H}_B)$ is

$$\mathcal{N}(X_A) = \sum_{a=1}^{d} \mathbf{M}_a X_A \mathbf{M}_a^{\dagger} \tag{2.22}$$

where $X_A \in \mathcal{L}(\mathcal{H}_A)$, $M_a \in \mathcal{L}(\mathcal{H}_A, \mathcal{H}_B)$, $\sum_{a=1}^{d} \mathbf{M}_a^{\dagger}\mathbf{M}_a = \mathbb{I}_{\dim(\mathcal{H}_A)}$, and $d \leq \dim(\mathcal{H}_A)\dim(\mathcal{H}_B)$. Here, $\dim(\mathcal{H}_*)$ refers to the dimension of the space $\mathcal{H}_*$.

Two common types of quantum channels are introduced next, which are widely used to simulate noise in quantum devices.

The first type is the *depolarizing channel*, which considers the scenario such that the information of the input state can be entirely lost with some probability.

Definition 2.5 (Depolarization Channel) Given an N-qubit quantum state $\rho \in \mathbb{C}^{2^N \times 2^N}$, the depolarization channel $\mathcal{N}_p$ acts on a 2^N-dimensional Hilbert space as follows

$$\mathcal{N}_p(\rho) = (1 - p)\rho + p\frac{\mathbb{I}_{2^N}}{2^N} , \tag{2.23}$$

where $\mathbb{I}_{2^N}/2^N$ refers to the maximally mixed state and p is a scalar representing the depolarization rate.

Example 2.5 (Single-Qubit State with Depolarization Channel) Consider a single-qubit pure state $\rho = |0\rangle \langle 0|$ with the density matrix

$$\rho = |0\rangle \langle 0| = \begin{bmatrix} 1 & 0 \\ 0 & 0 \end{bmatrix}. \tag{2.24}$$

When the depolarizing channel $\mathcal{N}_p$ acts on this state, the output is given by

$$\mathcal{N}_p(\rho) = (1 - p)\begin{bmatrix} 1 & 0 \\ 0 & 0 \end{bmatrix} + \frac{p}{2}\begin{bmatrix} 1 & 0 \\ 0 & 1 \end{bmatrix} = \begin{bmatrix} 1 - \frac{p}{2} & 0 \\ 0 & \frac{p}{2} \end{bmatrix}. \tag{2.25}$$

Therefore, the purity is inferred as

$$\mathrm{Tr}\left(\mathcal{N}_p^2(\rho)\right) = 1 - p + \frac{p^2}{2}. \tag{2.26}$$

When $p = 0$, the state remains pure and unchanged. When $0 < p \leq 1$, the state becomes a mixture of states $|0\rangle$ and $|1\rangle$ with $\mathrm{Tr}(\mathcal{N}_p^2(\rho)) < 1$. When $p = 1$, the state evolves into the maximally mixed state.

The second type is the *Pauli channel*, which serves as a dominant noise source in many computing architectures and as a practical model for analyzing error correction [8].

Definition 2.6 (Single-Qubit Pauli Channel) Given a quantum state $\rho \in \mathbb{C}^{2\times 2}$, the single-qubit Pauli channel $\mathcal{N}_{\mathbf{p}}$ acts on this state as follows

$$\mathcal{N}_{\mathbf{p}}(\rho) = p_I \rho + p_X \, \mathrm{X} \rho \, \mathrm{X} + p_Y \, \mathrm{Y} \rho \, \mathrm{Y} + p_Z \, \mathrm{Z} \rho \, \mathrm{Z} \,, \tag{2.27}$$

where $\mathbf{p} = (p_I, p_X, p_Y, p_Z)$ and $p_I + p_X + p_Y + p_Z = 1$.

Note that for a single-qubit system, the depolarization channel $\mathcal{N}_p$ is a special Pauli channel by setting $p_X = p_Y = p_Z = p$.

Example 2.6 (Single-Qubit State with Pauli Channel) Consider a single-qubit pure state $\rho = |0\rangle \langle 0|$ with the density matrix:

$$\rho = |0\rangle \langle 0| = \begin{bmatrix} 1 & 0 \\ 0 & 0 \end{bmatrix}. \tag{2.28}$$

When the Pauli channel $\mathcal{N}_{\mathbf{p}}$ acts on this state, the output is given by

$$\mathcal{N}_{\mathbf{p}}(\rho) = p_I \begin{bmatrix} 1 & 0 \\ 0 & 0 \end{bmatrix} + p_X \begin{bmatrix} 0 & 0 \\ 0 & 1 \end{bmatrix} + p_Y \begin{bmatrix} 0 & 0 \\ 0 & 1 \end{bmatrix} + p_Z \begin{bmatrix} 1 & 0 \\ 0 & 0 \end{bmatrix} \tag{2.29}$$

$$= \begin{bmatrix} p_I + p_Z & 0 \\ 0 & p_X + p_Y \end{bmatrix}. \tag{2.30}$$

Let us analyze three special cases for the probability vector $\mathbf{p} = (p_I, p_X, p_Y, p_Z)$:

- **Case 1**: If $p_X = p_Y = p_Z = p$, the Pauli channel reduces to the depolarization channel and the prepared state becomes

$$\mathcal{N}_{\mathbf{p}}(\rho) = \begin{bmatrix} 1 - 2p & 0 \\ 0 & 2p \end{bmatrix}. \tag{2.31}$$

- **Case 2**: If $p_Y = p_Z = 0$, the prepared state becomes

$$\mathcal{N}_{\mathbf{p}}(\rho) = \begin{bmatrix} 1 - p_X & 0 \\ 0 & p_X \end{bmatrix}. \tag{2.32}$$

In this scenario, the Pauli channel reduces to the *bit-flip channel*.

(continued)

Example 2.6 (continued)
- **Case 3**: For other values of **p**, the effect of the Pauli channel on the pure state $|0\rangle$ can be interpreted as a combination of the depolarizing channel and the bit-flip channel.

To generalize the single-qubit Pauli channel to a multi-qubit Pauli channel, the definition is extended to account for the action of Pauli operators on multiple qubits.

Definition 2.7 (Multi-qubit Pauli Channel) Given a quantum state $\rho \in \mathbb{C}^{2^N \times 2^N}$ for an N-qubit system, the multi-qubit Pauli channel $\mathcal{N}_\mathbf{p}$ acts as

$$\mathcal{N}_\mathbf{p}(\rho) = \sum_{P \in \mathcal{P}_N} p_P P \rho P^\dagger, \tag{2.33}$$

where $\mathcal{P}_N = \{I, X, Y, Z\}^{\otimes N}$ denotes the set of all tensor products of the N single-qubit Pauli operators and p_P is the probability of applying the Pauli operator P with $\sum_{P \in \mathcal{P}_N} p_P = 1$.

Remark
The multi-qubit Pauli channel considers the existence of correlated Pauli noise on different qubits. If each qubit only experiences independent single-qubit Pauli noise, the multi-qubit channel can be written as the tensor product of single-qubit Pauli channels:

$$\mathcal{N}_\mathbf{p}(\rho) = \otimes_{i=1}^{N} \mathcal{N}_{\mathbf{p}_i}(\rho), \tag{2.34}$$

where $\mathcal{N}_{\mathbf{p}_i}$ is the single-qubit Pauli channel acting on the i-th qubit with probabilities $\mathbf{p}_i = (p_I, p_X, p_Y, p_Z)$.

Given the motivation and definition of quantum channels, a natural question arises: *what is the relation between quantum channels and quantum gates?* It is straightforward to observe that a quantum gate is a special case of a quantum channel. Conversely, the evolution of a quantum state can be built from a unitary operation via isometric extension [7]. The following theorem demonstrates that any quantum channel arises from a unitary evolution on a larger Hilbert space.

Remark

The Choi–Kraus decomposition reveals that a unitary operator is a special case of a quantum channel. Specifically, when $d = 1$, the quantum channel reduces to

$$\mathcal{N}(X_A) = \mathbf{M}_1 X_A \mathbf{M}_1^{\dagger}, \tag{2.35}$$

where $\mathbf{M}_1$ is a unitary operator satisfying $\mathbf{M}_1^{\dagger}\mathbf{M}_1 = \mathbb{I}$. This highlights that all unitary operators are quantum channels, but not all quantum channels are unitary.

Theorem 2.1 ([7]) *Let $\mathcal{N}(\cdot) : \mathcal{L}(\mathcal{H}_A) \to \mathcal{L}(\mathcal{H}_B)$ be a quantum channel defined in Eq. (2.22). Let $\mathcal{H}_E$ be the Hilbert space of an auxiliary system. Denote the input state as ρ (i.e., a density operator $\rho \in \mathbb{C}^{dim(\mathcal{H}_A) \times dim(\mathcal{H}_A)}$). Then, there exists a unitary $U : \mathcal{L}(\mathcal{H}_A \otimes \mathcal{H}_E) \to \mathcal{L}(\mathcal{H}_B \otimes \mathcal{H}_E)$ and a normalized vector (i.e., a pure state) $|\varphi\rangle \in \mathbb{C}^{dim(\mathcal{H}_E)}$ such that*

$$\mathcal{N}(\rho) = \mathrm{Tr}_E\left(U(\rho \otimes |\varphi\rangle\langle\varphi|)U^{\dagger}\right), \tag{2.36}$$

where $Tr_E(\cdot)$ denotes the partial trace over the ancillary Hilbert space $\mathcal{H}_E$ and the dimension of $\mathcal{H}_E$ depends on the rank of the Kraus representation of $\mathcal{N}$.

Proof Sketch of Theorem 2.1 The system is extended to include an ancillary Hilbert space $\mathcal{H}_E$, representing the environment. The combined space $\mathcal{H}_A \otimes \mathcal{H}_E$ forms a closed physical system, whose evolution can be described by a unitary operator U acting on $\mathcal{H}_B \otimes \mathcal{H}_E$.

To find a feasible unitary U, the quantum channel $\mathcal{N}$ is rewritten using its isometric extension [7], i.e.,

$$\mathcal{N}(\rho) = \mathrm{Tr}_E\left(V\rho V^{\dagger}\right), \tag{2.37}$$

where $V : \mathcal{H}_A \to \mathcal{H}_B \otimes \mathcal{H}_E$ is an isometry operator embedding the input state into the larger Hilbert space. For simplicity, assume $\mathcal{H}_A = \mathcal{H}_B$. The isometry operator V can always be embedded into a unitary operator U acting on $\mathcal{H}_B \otimes \mathcal{H}_E$, ensuring that U captures the reversible evolution of the extended system.

Next, the input state ρ is augmented by introducing an ancillary state $|\varphi\rangle \in \mathcal{H}_E$, yielding the combined state $\rho \otimes |\varphi\rangle\langle\varphi|$. Substituting this augmented state and the unitary operator U into the isometric extension in Eq. (2.37) gives Eq. (2.36). Theorem 2.1 is thereby proven. $\qquad\qquad\square$

The translation between the unitary operation and the quantum channels described by Theorem 2.1 can be visually explained, as shown in Fig. 2.4. In

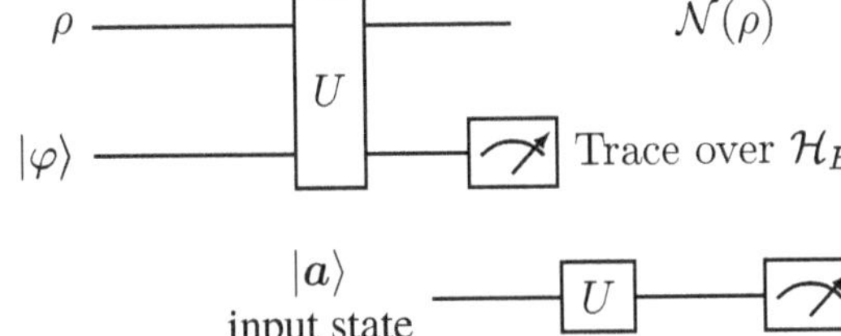

Fig. 2.4 The evolution of quantum states based on Theorem 2.1

Fig. 2.5 The quantum circuit diagram with measurement

this diagram, the first wire corresponds to the original input state ρ, while the second wire represents the initial state $|\varphi\rangle$ of the environment. To determine the output of the quantum channel $\mathcal{N}$ applied to ρ, an $\mathcal{N}$-induced unitary operation U is performed on the combined system, followed by a partial trace over the environment to discard its information.

2.2.2.3 Quantum Measurements

Besides quantum gates and channels, which change quantum states, measurement is another crucial operation in quantum circuits. Its goal is to get classical information from an evolved quantum state. Figure 2.5 shows a quantum circuit diagram: it depicts applying a unitary U to a single-qubit state $|a\rangle$, followed by a quantum measurement.

Quantum measurements fall into two main categories: *projective measurements* and *positive operator-valued measures* [1, 9].

Projective measurement, also known as von Neumann measurement, is described by a Hermitian operator $A = \sum_i \lambda_i |v_i\rangle \langle v_i|$, where $\{\lambda_i\}$ and $\{|v_i\rangle\}$ refer to the eigenvalues and eigenvectors of A, respectively. According to the Born rule [1], when the measurement operator $A \in \mathbb{C}^{2^N \times 2^N}$ is applied to an N-qubit state $|\Phi\rangle \in \mathbb{C}^{2^N}$, the probability of getting any eigenvalue from $\{\lambda_i\}$ is

$$\mathrm{Pr}(\lambda_i) = |\langle v_i|\Phi\rangle|^2 . \tag{2.38}$$

In the density operator representation, suppose that the state to be measured is $\rho \in \mathbb{C}^{2^N \times 2^N}$, the probability of measuring any one of the eigenvalues in $\{\lambda_i\}$ is

$$\mathrm{Pr}(\lambda_i) = \mathrm{Tr}(\rho|v_i\rangle \langle v_i|). \tag{2.39}$$

Define $\Pi_i = |v_i\rangle \langle v_i|$ as the i-th projective operator. The complete set of projective operators $\{\Pi_i\}$ has the following properties:

$$1)\ \Pi_i \Pi_j = \delta_{ij} \Pi_i;\ 2)\ \Pi_i^\dagger = \Pi_i;\ 3)\ \Pi_i^2 = \Pi_i;\ 4)\ \sum_i \Pi_i = \mathbb{I}_{2^N}. \tag{2.40}$$

One special set of projectors is $\Pi_i = |i\rangle \langle i|$ for $\forall i \in [2^N]$. This measures the probability of finding the system in the basis state $|i\rangle$. For example, given the single-qubit state $|\alpha\rangle$ in Eq. (2.2), the probability of measuring the computational basis state $|i\rangle$ is

$$\mathrm{Pr}(i) = |\langle v_i|\alpha\rangle|^2 = |\alpha_i|^2 . \tag{2.41}$$

The second type of quantum measurement is the *positive operator-valued measures (POVM)*. A POVM uses a set of positive operators $0 \preceq E_i$ satisfying $\sum_i E_i = \mathbb{I}$. Each positive operator E_i corresponds to a measurement outcome. Specifically, applying the measurement $\{E_m\}$ to the state $|\psi\rangle$, the probability of outcome i is given by

$$\mathrm{Pr}(i) = |\langle \psi|E_i|\psi\rangle|^2. \tag{2.42}$$

In the density operator representation, if the state to be measured is $\rho \in \mathbb{C}^{2^N \times 2^N}$, the probability of outcome i is

$$\mathrm{Pr}(i) = \mathrm{Tr}(\rho E_i). \tag{2.43}$$

The main difference between projective measurements and POVM elements is that POVM elements do not have to be orthogonal. Because of this, projective measurement is a special case of the generalized measurement (i.e., by setting $E_i = \Pi_i^\dagger \Pi_i$).

> **Remark**
> Here, we discuss what information can be accessed through quantum measurements, both in theory and in practice. To illustrate, imagine the computation result is the probability amplitude a_1 in the single-qubit state $|a\rangle = a_1|0\rangle + a_2|1\rangle$ in Eq. (2.2). To get a_1 as a classical value from this quantum state, the projective operator $\Pi_1 = |0\rangle \langle 0|$ is applied to this state. Quantum mechanics states that after each measurement, the state collapses. The measured outcome V_i acts as a binary random variable following a Bernoulli distribution $\mathrm{Ber}(a_i)$, i.e., $\mathrm{Pr}(V_i = 1) = a_1$ and $\mathrm{Pr}(V_i = 0) = 1 - a_1$. By applying the measurement Π_i to K copies of the state $|a\rangle$, the statistics are obtained. The sample mean is written as $\bar{a}_1 = \sum_{i=1}^{K} V_i/K$. The law of large numbers states that $\bar{a}_1 = a_1$ when $K \to \infty$. However, only the finite number of measurements K is allowed in practice. This leads to an estimation error.

2.3 Quantum Read-In and Read-Out Techniques

Quantum read-in and *read-out* describe how information moves between classical and quantum systems. These are fundamental steps in quantum machine learning (Fig. 1.1). They load data and extract results.

Quantum read-in and read-out are major roadblocks to using quantum computing for classical tasks. As emphasized in [10], quantum algorithms offer exponential speedups in certain problems, but these benefits are lost if data read-in and read-out are inefficient. Read-in means loading classical data into quantum systems, and read-out means extracting results from quantum systems. Specifically, quantum states are high dimensional, and measurement precision is limited. These factors often create overheads that grow quickly with problem size. These challenges highlight why optimizing quantum read-in and read-out is crucial to unlock the full potential of quantum computing. This section details quantum read-in and read-out methods, covering their basic concepts and several common algorithms.

2.3.1 Quantum Read-In

Quantum read-in is the process of encoding classical information into quantum systems that a quantum computer can manipulate. It's essentially a *classical-to-quantum mapping*. It bridges the gap, allowing us to use quantum algorithms to solve classical problems. This section introduces several common encoding methods: basis encoding, amplitude encoding, angle encoding, and quantum random access memory. Some easy-to-use demonstrations are provided in Sect. 2.5.

2.3.1.1 Basis Encoding

Basis encoding is a straightforward way to encode classical data that can be represented in binary form. Given a classical binary vector $x = (x_0, \ldots, x_i, \ldots, x_{N-1}) \in \{0, 1\}^N$, this method maps it directly to a quantum state:

$$|\psi\rangle = |x_0, \ldots, x_{N-1}\rangle. \tag{2.44}$$

This process requires N qubits to represent a binary vector of length N. To prepare the quantum state $|\psi\rangle$, we apply an X gate to each qubit if its corresponding bit value is 1. The overall quantum state preparation can be expressed as

$$|\psi\rangle = \bigotimes_{i=0}^{N-1} X^{x_i} |0\rangle^{\otimes N},$$

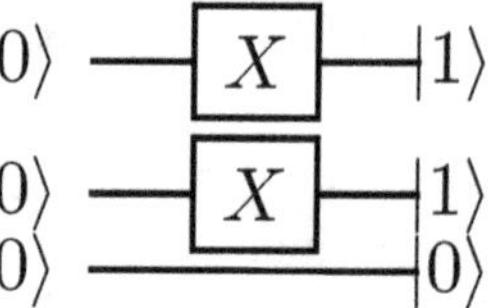

Fig. 2.6 Example of basis encoding for the integer 6

where $|0\rangle^{\otimes N}$ represents an initial state of all qubits set to $|0\rangle$ and X^{x_i} means applying the X gate to the i-th qubit only if $x_i = 1$.

> **Example 2.7 (Basis Encoding)** To encode the integer 6, its binary form is $x = (1, 1, 0)$. The corresponding quantum state is $|110\rangle$. This state can be implemented by applying X gates to the first and second qubits, as shown in Fig. 2.6.

2.3.1.2 Amplitude Encoding

Amplitude encoding maps classical data into the amplitudes of a quantum state. Given a vector $x = \left(x_0, \ldots, x_i, \ldots, x_{2^N-1}\right) \in \mathbb{C}^{2^N}$ containing complex values, the L_2 normalization is applied to obtain a normalized vector:

$$\hat{x} = \frac{x}{\|x\|_2}, \tag{2.45}$$

where $\|x\|_2$ is the Euclidean norm. This ensures that the normalized vector $\hat{x}$ satisfies $\sum_{i=0}^{2^N-1} |\hat{x}_i|^2 = 1$. The quantum state is then

$$|\psi\rangle = \sum_{i=0}^{2^N-1} \hat{x}_i |i\rangle \tag{2.46}$$

with $|i\rangle$ representing the N-qubit computational basis states.

> **Example 2.8 (Amplitude Encoding)** To encode a normalized vector $x = (x_0, x_1) \in \mathbb{C}^2$ as the quantum state $|\psi\rangle = x_0|0\rangle + x_1|1\rangle$, this can be achieved by applying a rotation gate $U = R_Y(\theta)$ to the initial state $|0\rangle$, where $\theta = 2\arccos(x_0)$.

Amplitude encoding can represent a very large vector $\left(\text{length } 2^N\right)$ with just N qubits, making it highly efficient in terms of qubit counts. However, preparing this quantum state requires constructing a unitary transformation U such that $|\psi\rangle = U|0\rangle^{\otimes N}$. Efficiently finding such transformations is challenging and an active research area (see Sect. 2.6 for more discussions).

2.3.1.3 Angle Encoding

Basis encoding and amplitude encoding are basic ways to map classical data to quantum states, but they have different resource needs. Basis encoding uses as many qubits as the data's binary length and needs few gate operations to prepare the state. In contrast, amplitude encoding is very qubit efficient, using only a logarithmic number of qubits for the data's size, but it needs many gate operations.

To address this limitation, an alternative is *angle encoding*. The core idea of angle encoding is to embed classical data into a quantum state through rotation angles.

Given a real-valued vector $x = (x_0, \ldots, x_i, \ldots, x_{N-1}) \in \mathbb{R}^N$, the encoded quantum state can be represented as

$$|\psi\rangle = \bigotimes_{i=0}^{N-1} R_\sigma(x_i)|0\rangle^{\otimes N} = \bigotimes_{i=0}^{N-1} \exp\left(-i\frac{x_i}{2}\sigma\right)|0\rangle^{\otimes N}, \tag{2.47}$$

where $\sigma \in \{X, Y, Z\}$ denotes a Pauli operator, as defined in Fig. 2.1. Since Pauli rotation gates are 2π-periodic, it is essential to scale each element x_i into the range $[0, \pi)$ to ensure that different values are encoded into distinct quantum states.

A key advantage of angle encoding is its ability to introduce nonlinearity. By mapping classical data into the parameters of quantum rotation gates, angle encoding uses trigonometric functions to naturally capture nonlinear relationships. This property is crucial in quantum machine learning because models need nonlinearity to learn complex patterns, like those that cannot be separated by a straight line.

2.3.1.4 Quantum Random Access Memory (QRAM)

Basis encoding, amplitude encoding, and angle encoding typically encode one data item at a time, making it hard to work with large, complex classical datasets. The QRAM [11], like classical RAM, can store, address, and access multiple quantum states at once.

QRAM consists of two types of qubits: data qubits for storing classical data and address qubits for addressing. Given a classical dataset $\mathcal{D} = \left\{x^{(j)}\right\}_{j=0}^{M-1}$ with M training examples, assume each data item is separately encoded into a quantum state $|x^{(j)}\rangle_d$ using one of the encoding methods above. The QRAM works like this: (1) First, prepare an N_a-qubit address register where $N_a = \lceil\log_2(M)\rceil$; (2) Then, link each data state $|x^{(j)}\rangle_d$ with corresponding address state $|j\rangle_a$. The entire dataset then forms a quantum state:

$$|\mathcal{D}\rangle = \sum_{j=0}^{M-1} \frac{1}{\sqrt{M}} |j\rangle_a |x^{(j)}\rangle_d.$$ (2.48)

Remark

The subscript d in $|x^{(j)}\rangle_d$ shows this quantum state is in the data register, unlike address qubits, which use the subscript a (e.g., $|j\rangle_a$). This convention helps to distinguish between the roles of data and address qubits in QRAM operations.

Example 2.9 (QRAM Encoding) Consider a dataset $\mathcal{D} = \{2, 3\}$. Using basis encoding, each sample is first turned into a two-qubit quantum state: $\{|10\rangle_d, |11\rangle_d\}$. Each data state then gets an address state, $|0\rangle_a$ for the first state $|10\rangle_d$ and $|1\rangle_a$ for the second state $|11\rangle_d$. The resulting QRAM-encoded state looks like this:

$$|\mathcal{D}\rangle = \frac{1}{\sqrt{2}} \left(|0\rangle_a |10\rangle_d + |1\rangle_a |11\rangle_d\right).$$ (2.49)

The corresponding quantum circuit for implementing this state is shown in Fig. 2.7.

QRAM allows the dataset $\mathcal{D}$ to be stored in a coherent quantum superposition, enabling simultaneous access to all data items through the entanglement of address and data qubits. While QRAM is theoretically powerful, its practical implementation remains a significant challenge due to the need for a large number of qubits and quantum operations (see Sect. 2.6 for the discussion).

Fig. 2.7 Example of QRAM encoding for the dataset $\mathcal{D} = \{2, 3\}$

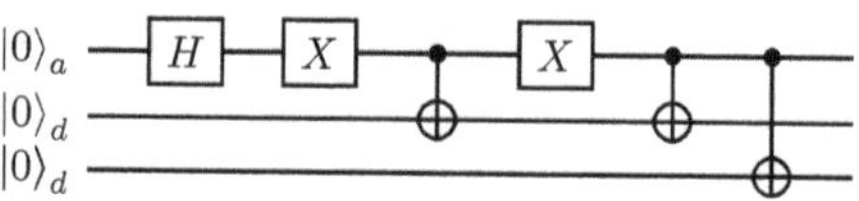

2.3.2 Quantum Read-Out Methods

Quantum read-out translates the quantum state from a computation into classical data. This allows for further processing, interpretation, or optimization using classical systems. It's the opposite of quantum read-in, acting as a quantum-to-classical mapping.

Depending on how much information is extracted, quantum read-out methods generally fall into two types: *full information and partial information read-out methods*. These methods allow for customized read-out processes that fit the needs of various quantum applications, including tomography, optimization, and machine learning tasks.

2.3.2.1 Full Information Read-Out Methods

Full information read-out aims to completely reconstruct the quantum state. This helps us fully understand how the quantum system behaves. The most common way to do this is through quantum state tomography (QST) [12].

QST involves taking quantum measurements, collecting statistics, and then using classical computers to reconstruct the quantum state. In what follows, two reconstruction techniques broadly used in QST, i.e., linear inversion [13] and maximum likelihood estimation (MLE) [14], are introduced.

QST with Linear Inversion Linear inversion is a direct way to reconstruct a quantum state from measurement data by solving linear equations. Let ρ be the explored quantum state and $\{E_i\}$ be a set of measurements. According to the Born rule, the probability of measurement outcome i is given by

$$\Pr(E_i|\rho) = \mathrm{Tr}(\rho E_i). \tag{2.50}$$

In practice, $\Pr(E_i|\rho)$ is not directly accessible but is approximated by the frequency p_i of measurement outcome i over multiple measurements. By the law of large numbers, as the number of measurements increases, p_i converges to the true probability $\Pr(E_i|\rho)$. Collecting measurements across all bases, there is a linear system:

$$\begin{bmatrix} \mathrm{Tr}(\rho E_0) \\ \mathrm{Tr}(\rho E_1) \\ \vdots \end{bmatrix} = \begin{bmatrix} \mathbf{E}_0^\dagger \cdot \boldsymbol{\rho} \\ \mathbf{E}_1^\dagger \cdot \boldsymbol{\rho} \\ \vdots \end{bmatrix} = A\boldsymbol{\rho} \approx \boldsymbol{p} = \begin{bmatrix} p_0 \\ p_1 \\ \vdots \end{bmatrix}, \tag{2.51}$$

where $\mathbf{E}$ and $\boldsymbol{\rho}$ refer to the vector representations of matrices E_i and ρ, respectively. The vector representation of a matrix is obtained by stacking its columns into a single-column vector. For example, the vector form of a 2×2 identity matrix is $\mathbb{I}_2 = [1, 0, 0, 1]^T$. The matrix A is constructed such that each row corresponds to the vector representation of the measurement operator, i.e., $A = \left[\mathbf{E}_0^\dagger; \mathbf{E}_1^\dagger; \dots \right]$. The vector $\boldsymbol{p}$ contains the measured frequencies p_i.

If the measurements are tomographically complete (meaning $\{E_i\}$ forms a basis for the system's Hilbert space), we can reconstruct the state ρ by solving these linear equations:

$$\rho = \left(A^T A\right)^{-1} A^T \boldsymbol{p}. \tag{2.52}$$

A common strategy is to use Pauli operators as measurement bases $\{E_i\}$. The density matrix ρ for an N-qubit system can be written using the Pauli basis as

$$\rho = \frac{1}{2^N} \sum_{i=0}^{4^N-1} c_i P_i, \quad c_i \in \mathbb{R}, \quad P_i \in \{I, X, Y, Z\}^{\otimes N}. \tag{2.53}$$

The coefficients c_i represent projections of ρ onto the Pauli basis, calculated as

$$c_i = \mathrm{Tr}(\rho P_i). \tag{2.54}$$

To fully reconstruct ρ, the quantum state must theoretically be measured in all $4^N - 1$ Pauli bases to estimate each c_i.

Remark

The Pauli basis includes four Hermitian matrices: I, X, Y, and Z, as introduced in Fig. 2.1. These operators form a complete basis for the space of 2×2 complex matrices. For N-qubit systems, the tensor products of these single-qubit operators span the space of $2^N \times 2^N$ complex matrix. This makes the Pauli basis essential for representing quantum states, observables, and their transformations.

A key drawback of linear inversion is that it doesn't guarantee a valid density matrix. The estimated quantum state might not have properties like positive semi-definiteness (Definition 2.2), especially with limited measurements.

Maximum Likelihood Estimation (MLE) To ensure physical constraints on the quantum state during reconstruction, MLE is introduced. MLE reconstructs ρ by maximizing the likelihood of observing the measurement outcomes. It does this while ensuring ρ is Hermitian, is positive semi-definite, and has a trace of one. The likelihood function is

$$L(\rho) = \prod_i \mathrm{Tr}(\rho E_i)^{p_i}. \tag{2.55}$$

Reconstructing ρ then becomes solving this optimization problem:

$$\arg\max_{\rho'} L\left(\rho'\right), \quad \text{s.t.} \quad \rho' \succeq 0, \quad \rho' = \rho'^{\dagger}, \quad \text{Tr}\left(\rho'\right) = 1. \tag{2.56}$$

Solving this usually needs iterative numerical optimization, which can be computationally intensive.

> **Remark**
> A common challenge across all quantum state tomography (QST) methods, including linear inversion and MLE, is the *exponential computational cost* with respect to the number of qubits. Specifically, the number of parameters needed for reconstruction grows exponentially with the system size. This makes QST methods practical only for systems with a small number of qubits. This limitation highlights the need for scalable ways to characterize quantum states in larger systems.

2.3.2.2 Partial Information Read-Out Methods

Partial information read-out methods extract specific, useful information from a quantum state without needing to reconstruct its entire density matrix. This approach enables efficient characterization and analysis of large-qubit systems. Current partial read-out techniques generally fall into three categories based on the type of information collected: sampling, expectation value estimation, and shadow tomography.

Sampling Sampling involves repeatedly measuring the quantum state in the computational basis to estimate the probability distribution over bit-strings. Given a state $|\psi\rangle$, the probability of observing a specific computational basis $|i\rangle$ is given by

$$\Pr(i) = |\langle\psi|i\rangle|^2. \tag{2.57}$$

The frequency of each outcome from repeated measurements provides an estimate of $\Pr(i)$. Sampling is particularly useful in the following applications:

- Sampling over complicated distributions. Quantum states can represent complex probability distributions that are difficult to sample classically. Quantum sampling allows efficient exploration of these distributions for specific applications, such as probabilistic modeling and Markov chain Monte Carlo [15].
- Optimization problems. Sampling helps find high-probability bit-strings in quantum algorithms like the Quantum Approximate Optimization Algorithm [16] and

Grover search [17]. These sampled bit-strings often represent optimal or near-optimal solutions.

- Verification. Sampling facilitates the comparison of a quantum circuit's output with theoretical expectations or desired distributions, helping to verify the quantum systems [2, 18, 19].

Expectation Value Estimation For many quantum computation problems, such as in quantum chemistry and many-body physics, the computation result is the estimated expectation value of certain observables on the evolved quantum state [20, 21].

An observable $O \in \mathbb{C}^{2^N \times 2^N}$ mentioned here is a Hermitian operator that represents a measurable physical quantity. For an N-qubit system, O can be expressed in terms of a Pauli basis expansion, i.e.,

$$O = \sum_{i=1}^{4^N} \alpha_i P_i, \quad P_i \in \{\mathbb{I}_2, X, Y, Z\}^{\otimes N}, \quad \alpha_i \in \mathbb{R}. \tag{2.58}$$

where P_i denotes the i-th N-qubit Pauli string.

The expectation value of an observable O with respect to an N-qubit state ρ is

$$\langle O \rangle = \mathrm{Tr}(\rho O). \tag{2.59}$$

Substituting the Pauli expansion of O, the expectation value is expressed as the weighted sum of the expectation values of each Pauli basis term due to the linearity of the trace operation, i.e.,

$$\langle O \rangle = \sum_{i=1}^{4^N} \alpha_i \, \mathrm{Tr}(\rho P_i) \equiv \sum_{i=1}^{4^N} \alpha_i \, \langle P_i \rangle . \tag{2.60}$$

To estimate the expectation value of each individual Pauli term P_i, the quantum state ρ must be measured on the basis of the eigenstates of P_i. The measurement outcome is then associated with the corresponding eigenvalue of P_i. Notably, the eigenstates and eigenvalues of P_i can be derived from the eigenstates and eigenvalues of its constituent single-qubit Pauli operators P_{ij}.

- Eigenvalues. The eigenvalues of P_i are the product of the eigenvalues of each single-qubit Pauli operator P_{ij}, i.e., $P_i = \otimes_{j=1}^{N} P_{ij}$. For example, if the eigenvalues of P_{ij} are ± 1, then the eigenvalues of P_i are products of these individual eigenvalues and remain in $\{\pm 1\}$.
- Eigenstates. The eigenstates of P_i are the tensor products of the eigenstates of the single-qubit Pauli operators P_{ij}. If $|\lambda_{ijk}\rangle$ is one of the eigenstate of P_{ij}, then the corresponding eigenstate of P_i is $\bigotimes_{j=1}^{N} |\lambda_{ijk}\rangle$.

This structure allows P_i to be analyzed in terms of its simpler single-qubit components, significantly simplifying the process of determining the measurement basis for expectation value estimation. By repeating the measurements M times and obtaining the corresponding measurement results $\{r_j\}_{j=1}^{M}$, the statistical value of $\langle P_i \rangle$ can be estimated by

$$\langle \hat{P}_i \rangle = \frac{1}{M} \sum_{j=1}^{M} r_j. \tag{2.61}$$

The expectation value of the observable O is therefore statistically estimated by $\langle \hat{O} \rangle = \sum_{i=0}^{4^N - 1} \alpha_i \langle \hat{P}_i \rangle$.

Remark

A key step in the process is to measure the quantum system in the basis of the eigenstates of P_i. If P_i is diagonal in the computational basis (e.g., a tensor product of Pauli-Z operators), the state can be directly measured without additional operations. Otherwise (e.g., for Pauli-X or Pauli-Y operators), a unitary transformation must be applied to rotate the quantum state into the desired basis. Specifically, when measuring in the Pauli-X basis (i.e., $|+\rangle$ and $|-\rangle$), a Hadamard gate H is applied to the state ρ, i.e.,

$$\rho' = H\rho H. \tag{2.62}$$

When measuring in the Pauli-Y basis (i.e., $\frac{|0\rangle + i|1\rangle}{\sqrt{2}}$ and $\frac{|0\rangle - i|1\rangle}{\sqrt{2}}$), a phase gate $S = \sqrt{Z}$ followed by a Hadamard gate H is applied, i.e.,

$$\rho' = S^{\dagger} H \rho H S. \tag{2.63}$$

Measuring the state ρ' in the computational basis is equivalent to measuring the state ρ in the corresponding Pauli basis.

Shadow Tomography Full QST needs an exponential number of quantum state copies, making it impractical for systems with more than a few qubits. Instead of fully reconstructing the density matrix, shadow tomography [22] efficiently extracts specific properties of a quantum state, such as the expectation values of many observables.

Definition 2.8 (Shadow Tomography, [22]) Given an unknown D-dimensional quantum state ρ, as well as M observables $O_1, \ldots, O_M$, output real numbers $b_1, \ldots, b_M$ such that $|b_i - \mathrm{Tr}(O_i \rho)| \leq \epsilon$ for all i, with success probability at least $1 - \delta$. Do this via a measurement of $\rho^{\otimes k}$, where $k = k(D, M, \epsilon, \delta)$ is as small as possible.

Aaronson [22] proved that the shadow tomography problem can be solved using a *polylogarithmic* number of copies of states in terms of the dimension D and number M of observables. This result demonstrates that it is possible to estimate the expectation values of exponentially many observables for a quantum state of exponential dimension using only a polynomial number of measurements.

The core idea of shadow tomography is to create a compact classical representation, or "shadow," of a quantum state. This shadow holds enough information to estimate many of the state's properties. Building on this concept, [23] proposed a more practical and efficient approach, termed *classical shadow*, which uses randomized measurements to construct this classical representation. The classical shadow approach consists of the following steps:

1. Randomized measurements. Perform random unitary transformations on the quantum state and measure the transformed state in the computational basis. These random transformations can be drawn from specific ensembles, such as Clifford gates or local random rotations, which ensure that the measurement outcomes capture the essential properties of the quantum state.
2. Classical shadow construction. Using the measurement results, construct a classical shadow of the quantum state. This compact representation encodes the quantum state in a way that allows for the efficient estimation of properties.
3. Property estimation. Use the classical shadow to compute the desired properties of the quantum state, such as expectation values of specific observables, subsystem entropies, or fidelities with known states.

Shadow tomography needs exponentially fewer measurements than full quantum state tomography. This makes it a practical solution for large-scale quantum systems. Moreover, the shadow of a quantum state serves as a versatile representation, enabling the efficient estimation of various properties such as expectation values, entanglement measures, and subsystem correlations.

2.4 Quantum Linear Algebra

We next introduce quantum linear algebra, a potent toolbox for designing various FTQC-based algorithms introduced in Sect. 1.2.2. For clarity, the definition of block encoding is presented in Sect. 2.4.1, detailing how to implement a matrix on a quantum computer. Based on this, some basic arithmetic rules for block encodings are introduced in Sect. 2.4.2, including multiplication, linear combination, and the Hadamard product. Finally, in Sect. 2.4.3, the quantum singular value transformation method is introduced, which enables one to implement functions onto singular values of block-encoded matrices.

2.4.1 Block Encoding

For many computational problems, such as solving linear equations, it is often necessary to deal with a non-unitary matrix A. However, remember that quantum gates as discussed in Sect. 2.2 are unitaries. Therefore, if we want to solve these problems on quantum computers, it is essential to consider how to encode the matrix A into a unitary. This challenge can be addressed by the block encoding technique.

Definition 2.9 (Block Encoding, [24]) Suppose that A is an N-qubit operator, $\alpha, \varepsilon \geq 0$ and $a \in \mathbb{N}$. Then, the $(a + N)$-qubit unitary U is said to be an (α, a, ε)-block encoding of A if

$$\|A - \alpha(\langle 0|^{\otimes a} \otimes \mathbb{I}_{2^N})U(|0\rangle^{\otimes a} \otimes \mathbb{I}_{2^N})\| \leq \varepsilon. \tag{2.64}$$

Here, $\|\cdot\|$ represents the spectral norm, i.e., the largest singular value of the matrix.

The circuit implementation of the block encoding is illustrated in Fig. 2.8. The scaled matrix A/α interacts with the state $|\psi\rangle$ if the first qubit registers are measured as $|0\rangle$. By definition, there is $\alpha \geq \|A\|$ and any unitary U is an $(1, 0, 0)$-block encoding of itself.

Fact 2.2 (Block Encoding via the Linear Combination of Unitaries (LCU) Method, [24]) *Suppose that A can be written in the form*

$$A = \sum_k \alpha_k U_k, \tag{2.65}$$

where $\{\alpha_k\}$ are real numbers and U_k are some easily prepared unitaries such as Pauli strings. Then, the LCU method allows us to have the access to two unitaries, i.e.,

$$U_{\mathrm{SEL}} = \sum_k |k\rangle\langle k| \otimes U_k, \tag{2.66}$$

$$U_{\mathrm{PREP}} : |0\rangle \rightarrow \frac{1}{\sqrt{\|\alpha\|_1}} \sum_k \sqrt{\alpha_k}|k\rangle, \tag{2.67}$$

where $\alpha = (\alpha_1, \alpha_2, \dots)$.

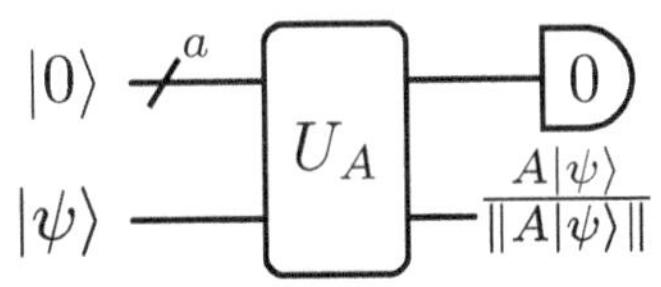

Fig. 2.8 Quantum circuit for block encoding

After simple mathematical analysis, one can obtain $U = (U_{\mathrm{PREP}}^{\dagger} \otimes \mathbb{I}_{2^N}) U_{\mathrm{SEL}} (U_{\mathrm{PREP}} \otimes \mathbb{I}_{2^N})$ is a $(\|\boldsymbol{\alpha}\|_1, m, 0)$-block encoding of A. Here, $\mathbb{I}_{2^N}$ is the identity operator of N-qubit size and $\| \cdot \|_1$ denotes the ℓ_1 norm of a given vector.

Similar to the definition of block encoding, the state preparation encoding can be defined.

Definition 2.10 (State Preparation Encoding [25]) We say a unitary U_ψ is an (α, a, ϵ)-state-encoding of an N-qubit quantum state $|\psi\rangle$ if

$$\| |\psi\rangle - \alpha(\langle 0^a | \otimes I) U_\psi |0^{a+N}\rangle \|_\infty \le \epsilon, \tag{2.68}$$

where $\| \cdot \|_\infty$ denotes the infinity norm of the given vector.

More straightforwardly, the (α, a, ϵ)-state-encoding U_ψ prepares the state:

$$U_\psi |0\rangle |0\rangle = \frac{1}{\alpha} |0\rangle |\psi'\rangle + \sqrt{1 - \alpha^2} |1\rangle |\mathrm{bad}\rangle,$$

where $\| |\psi'\rangle - |\psi\rangle \|_\infty \le \epsilon$ and $|\mathrm{bad}\rangle$ is an arbitrary quantum state. One can further prepare the state $|\psi'\rangle$ by using $O(\alpha)$ times of amplitude amplification [26]. The state preparation encoding can be understood as a specific case of the block encoding, i.e., it is the block encoding of a $\mathbb{C}^{2^N \times 1}$ matrix.

2.4.2   Basic Arithmetic for Block Encodings

Now, we introduce some arithmetic rules for block encoding unitaries. The following two facts describe the product and linear combination rules of block encoding unitaries, respectively.

Fact 2.3 (Product of Block Encoding, [24]) *If U is an (α, a, δ)-block encoding of an N-qubit operator A, and V is a (β, b, ε)-block encoding of an N-qubit operator B, then $(\mathbb{I}_{2^b} \otimes U)(\mathbb{I}_{2^a} \otimes V)$ is an $(\alpha\beta, a + b, \alpha\varepsilon + \beta\delta)$-block encoding of AB. Here, $\mathbb{I}_{2^a}$ is the identity operator of a-qubit size.*

Fact 2.4 (Linear Combination of Block Encoding, [24]) *Let $A = \sum_k x_k A_k$ be an s-qubit operator with $\beta \ge \|\mathbf{x}\|_1$ and $\varepsilon_1 > 0$, where $\mathbf{x}$ is the vector of coefficients. Suppose the access to*

$$P_L |0\rangle = \sum_k c_k |k\rangle, \tag{2.69}$$

$$P_R |0\rangle = \sum_k d_k |k\rangle, \tag{2.70}$$

$$W = \sum_k |k\rangle\langle k| \otimes U_k + \left(\left(\mathcal{I}_s - \sum_k |k\rangle\langle k| \right) \otimes \mathcal{I}_a \otimes \mathcal{I}_b \right), \qquad (2.71)$$

where $\sum_k |\beta c_k^* d_k - x_k| \leq \varepsilon_1$ and U_k is an $(\alpha, a, \varepsilon_2)$-block encoding of A_k. Then, an $(\alpha\beta, a + b, \alpha\varepsilon_1 + \beta\varepsilon_2)$-block encoding of A can be implemented by using one time of W, P_L, and P_R.

These results can be verified via direct computation. Another arithmetic rule broadly employed in quantum machine learning is the Hadamard product, a.k.a, the element-wise product. The following lemma exhibits how to achieve this operation via the block encoding framework.

Lemma 2.1 (Hadamard Product of the Block Encoding Unitaries, [25]) *With $N \in \mathbb{N}$, consider two matrices $A, B \in \mathbb{C}^{2^N \times 2^N}$, and assume that an (α, a, δ)-encoding U_A of matrix A and (β, b, ϵ)-encoding U_B of matrix B are accessible, then an $(\alpha\beta, a + b + N, \alpha\epsilon + \beta\delta)$-encoding of matrix $A \circ B$ corresponding to the Hadamard product of A and B can be constructed.*

Proof Sketch of Lemma 2.1 For simplicity, we only consider the perfect case, i.e., no errors. Refer to Ref. [25] for the proof details under the more general cases.

The intuition for achieving the Hadamard product is that all the needed elements can be found in the tensor product, i.e.,

$$\left(\langle 0^{a+b}| \otimes \mathbb{I}_{2^{2N}} \right) \left(\mathbb{I}_{2^b} \otimes U_A \otimes \mathbb{I}_{2^N} \right) \left(\mathbb{I}_{2^a} \otimes U_B \otimes \mathbb{I}_{2^N} \right) \left(|0^{a+b}\rangle \otimes \mathbb{I}_{2^{2N}} \right) \qquad (2.72)$$

$$= \frac{A \otimes B}{\alpha\beta}.$$

To this end, the question is reduced to finding proper permutation unitaries that can shift the required elements to the correct positions to achieve the Hadamard product. Denote $P' = \sum_{i=0}^{d-1} |i\rangle\langle i| \otimes |0\rangle\langle i|$. As proved by Zhao et al. [27], the tensor product of A and B can be reformulated to the Hadamard product via P, i.e.,

$$P'(A \otimes B)P'^\dagger = (A \circ B) \otimes |0\rangle\langle 0|.$$

However, P' is not a unitary. Instead, we consider $P = \sum_{i,j=0}^{d-1} |i\rangle\langle i| \otimes |i \oplus j\rangle\langle j|$, which can be easily constructed by using N CNOT gates, i.e., one CNOT gate between each pair of qubits consisting of one qubit from the first register and the corresponding qubit from the second register. By direct computation, it can be shown

$$\left(\mathbb{I}_{2^N} \otimes \langle 0^N| \right) P \left(A \otimes B \right) P^\dagger \left(\mathbb{I}_{2^N} \otimes |0^N\rangle \right) = A \circ B. \qquad (2.73)$$

Therefore, by direct computation, one can verify that $\left(P \otimes \mathbb{I}_{2^{a+b}} \right) \left(\mathbb{I}_{2^b} \otimes U_A \otimes \mathbb{I}_{2^N} \right) \left(\mathbb{I}_{2^a} \otimes U_B \otimes \mathbb{I}_{2^N} \right) \left(P^\dagger \otimes \mathbb{I}_{2^{a+b}} \right)$ is the desired block encoding. $\qquad\square$

2.4.3 Quantum Singular Value Transformation

Having established how to implement matrices on quantum computers and perform arithmetic operations among them, the next step is to introduce methods for implementing matrix functions on quantum computers. The method is called the quantum singular value transformation (QSVT), which is a powerful framework that can unify most known quantum algorithms.

In machine learning applications, we primarily work with real matrices. The computational cost of QSVT for real matrices is summarized in the following theorem. For a matrix A, consider its singular value decomposition $A = \sum_i \sigma_i |\psi_i\rangle\langle\phi_i|$. Given a function $P(x)$, $P^{(SV)}(A)$ is used to represent $P^{(SV)}(A) = \sum_i P(\sigma_i)|\psi_i\rangle\langle\phi_i|$.

Fact 2.5 (Quantum Singular Value Transformation—Real Matrix Case, [24])
Suppose that U_A is an (α, a, ε)-block encoding of a real matrix A. If $\delta \geq 0$ and $P : \mathbb{R} \to \mathbb{C}$ is a d-degree polynomial satisfying that

$$\text{for all } x \in [-1, 1] : |P(x)| \leq \frac{1}{4}, \tag{2.74}$$

then a quantum circuit $\tilde{U}$ exists. This circuit is a $(1, a + 3, 4d\sqrt{\varepsilon/\alpha} + \delta)$-block encoding of $P^{(SV)}(A/\alpha)$ and involves d applications of U_A and $U_A^\dagger$ gates. Further, the circuit's description can be classically computed in $O(\mathrm{poly}(d, \log(1/\delta)))$ time.

If the block-encoded matrix A is Hermitian, the singular value transformation is equivalent to the eigenvalue transformation. In this case, the matrix function can be directly implemented using QSVT.

The following section introduces several applications of QSVT. While QSVT has many important applications, this book focuses on those relevant to machine learning tasks. The first application is matrix inversion, widely used in traditional machine learning methods like principal component analysis. For a general matrix, this typically refers to implementing the Moore-Penrose pseudoinverse, i.e., the inverse of all singular values.

Lemma 2.2 (Matrix Inversion, Simplified [24]) *Let U_A be a $(1, a, 0)$-block encoding of matrix A. Further, for simplicity, assume the nonzero singular values of A are lower bounded by $\delta > 0$. Let $0 \leq \epsilon \leq \delta \leq \frac{1}{2}$. One can construct a $(1/\delta, a + 2, \epsilon)$-block encoding of A^{-1} by using $\tilde{O}(\frac{1}{\delta}\log(\frac{1}{\epsilon}))$ times of U_A and $U_A^\dagger$.*

Proof Sketch of Lemma 2.2 This can be achieved by finding a good polynomial approximation for the function $1/x$. While such a polynomial cannot be found for the entire interval $[-1, 1]$, it does exist on the interval $[-1, -\delta] \cup [\delta, 1]$ for some $\delta > 0$. □

The second application of QSVT is nonlinear amplitude transformation. As mentioned in Sect. 2.3.1, classical data can be encoded into quantum states in several ways. Here, the employed methodology is the amplitude encoding case described in Sect. 2.3.1.2, especially for the real amplitudes. The nonlinear transformation is achieved by combining diagonal block encoding with QSVT.

Fact 2.6 (Diagonal Block Encoding of Amplitudes, [28, 29]) *Given a state preparation unitary U_ψ of an N-qubit state $|\psi\rangle = \sum_{j=1}^{2^N} \psi_j |j\rangle$, where $\{\psi_j\}$ are real, $\|\psi\|_2 = 1$, one can construct an $(1, N+2, \epsilon)$-encoding of the diagonal matrix $A = \mathrm{diag}(\psi_1, \ldots, \psi_d)$ with $O(N)$ circuit depth and $O(1)$ times of controlled-U and controlled-$U^\dagger$.*

As a straightforward generalization, one can replace the state preparation unitary with the general state preparation encoding, as mentioned in Definition 2.10. By creating the block encoding of amplitudes, we can implement many functions onto these amplitudes using QSVT. A direct application is performing the neural network on the quantum computer, as will be detailed in Sect. 5.2.

2.5 Code Demonstration

This section provides code implementations for key techniques introduced earlier, including quantum read-in strategies and block encoding. These examples offer readers an opportunity to practice and deepen their understanding.

2.5.1 Read-In Implementations

This subsection demonstrates toy examples of implementing data encoding methods in quantum computing, as discussed in earlier sections. Specifically, the basis encoding, amplitude encoding, and angle encoding from Sect. 2.3.1.3 are covered. These examples aim to provide readers with hands-on experience in applying quantum data encoding techniques.

2.5.1.1 Basis Encoding

PennyLane provides built-in support for basis encoding through its "BasisEmbedding" function. Below is Python code demonstrating the basis encoding for the integer 6.

```python
import pennylane as qml

dev = qml.device("default.qubit", range(3))
@qml.qnode(dev)
def circuit(x):
        qml.BasisEmbedding(x, range(3))
        return qml.state()

# Call the function
circuit(6)
```

2.5.1.2 Amplitude Encoding

PennyLane offers built-in support for amplitude encoding via the "AmplitudeEm-bedding" function. Below is a Python example demonstrating amplitude encoding for a randomly generated complex vector.

```python
import pennylane as qml
import numpy as np

# Number of qubits
n_qubits = 8

# Define a quantum device with 8 qubits
dev = qml.device("default.qubit", wires=n_qubits)

@qml.qnode(dev)
def circuit(x):
    qml.AmplitudeEmbedding(features=x, wires=range(n_qubits),
        normalize=True, pad_with=0.)
    return qml.state()

# Generate a random complex vector of length 2^n_qubits
x_real = np.random.normal(loc=0, scale=1.0, size=2**n_qubits)
x_imag = np.random.normal(loc=0, scale=1.0, size=2**n_qubits)
x = x_real + 1j * x_imag

# Execute the circuit to encode the vector as a quantum state
circuit(x)
```

2.5.1.3 Angle Encoding

PennyLane provides built-in support for angle encoding via the "AngleEmbedding" function. Below is a Python example demonstrating angle encoding for a randomly generated real vector.

```python
import pennylane as qml
import numpy as np

# Number of qubits
n_qubits = 8

# Define a quantum device with 8 qubits
dev = qml.device("default.qubit", wires=n_qubits)


@qml.qnode(dev)
def circuit(x):
    qml.AngleEmbedding(features=x, wires=range(n_qubits),
        rotation="X")
    return qml.state()

# Generate a random real vector of length n_qubits
x = np.random.uniform(0, np.pi, (n_qubits))

# Execute the circuit to encode the vector as a quantum state
circuit(x)
```

2.5.2   Block Encoding

An example of constructing a block encoding is provided here. The block encoding
is constructed via the linear combination, as described in Fact 2.2. PennyLane is
used to maintain consistency, but there are also many other available platforms.
Note that Pauli decomposition is time-consuming (for an N-qubit matrix, it takes
$O(N4^N)$ time), so it's not recommended to try this method with large matrices.

```python
import numpy as np
import pennylane as qml
import matplotlib.pyplot as plt

a = 0.36
b = 0.64

# matrix to be decomposed
A = np.array(
    [[a,  0, 0,  b],
     [0, -a, b,  0],
     [0,  b, a,  0],
     [b,  0, 0, -a]]
)

# decompose the matrix into sum of Pauli strings
LCU = qml.pauli_decompose(A)
LCU_coeffs, LCU_ops = LCU.terms()

# normalized square roots of coefficients
```

```python
21  alphas = (np.sqrt(LCU_coeffs) / np.linalg.norm(np.sqrt(
        LCU_coeffs)))
22
23  dev = qml.device("default.qubit", wires=3)
24
25  # unitaries
26  ops = LCU_ops
27  # relabeling wires: 0 --> 1, and 1 --> 2
28  unitaries = [qml.map_wires(op, {0: 1, 1: 2}) for op in ops]
29
30  @qml.qnode(dev)
31  def lcu_circuit():  # block_encode
32      # PREP
33      qml.StatePrep(alphas, wires=0)
34
35      # SEL
36      qml.Select(unitaries, control=0)
37
38      # PREP_dagger
39      qml.adjoint(qml.StatePrep(alphas, wires=0))
40      return qml.state()
41
42  print(np.real(np.round(output_matrix,2)))
```

2.6 Bibliographic Remarks

This chapter concludes by discussing recent advancements in efficiently implementing fundamental quantum computing components. For clarity, a brief overview of advanced quantum read-in and read-out methods is provided, as these are crucial for efficiently loading and extracting classical data in quantum machine learning pipelines. The latest progress in quantum linear algebra is also reviewed.

2.6.1 Advanced Quantum Read-In Methods

While conventional read-in methods can encode classical data into quantum computers, they typically face two key challenges that limit their broad use in practical learning problems. To address these limitations, initial efforts have focused on developing more advanced quantum read-in methods.

Challenge I: High demand for quantum resources. Encoding methods like amplitude encoding and basis encoding presented in Sect. 2.3.2 generally suffer from high quantum resource requirements. While amplitude encoding is highly compact in terms of qubit requirements, it demands an exponential number of quantum gates relative to data size to prepare an exact amplitude-encoded state. In contrast, basis encoding requires a large number of qubits proportional to the input size, even though it can be implemented with a small number of quantum gates. The high demand for either quantum gates or qubit counts makes these basic encoding strategies infeasible for practical use.

Challenge II: Insufficient nonlinearity. While quantum mechanics is inherently linear, most practical machine learning models require nonlinearity to capture complex data patterns effectively. Conventional encoding methods like angle encoding introduce some nonlinearity. However, their representational power remains limited due to the linear nature of quantum operations and shallow circuit depth.

For Challenge I, a practical alternative is the approximate amplitude encoding (AAE) [30]. Instead of implementing exact amplitude encoding, AAE trains a parameterized quantum circuit with a constrained depth to approximate the desired quantum state with high fidelity. The training process optimizes the fidelity between the target state and the approximate state, ensuring that the representation error remains within a small bound.

For Challenge II, techniques like *data re-uploading* [31] have been developed. *Data re-uploading* involves feeding the same classical data into the quantum circuit multiple times, interspersed with trainable quantum operations. By alternating data encoding with trainable transformations, this approach allows the quantum model to capture nonlinear relationships more effectively without additional qubits. Additionally, neural quantum embedding [32] has been proposed. This method leverages classical deep learning techniques to learn optimal quantum embeddings, effectively separating nonlinearly separable data classes.

To address both Challenges I and II, hybrid encoding strategies have been introduced to leverage the respective advantages of each encoding method. For instance, basis-amplitude encoding combines basis encoding for discrete random variables with amplitude encoding for high-precision probabilities. This effectively encodes both categorical and continuous features without additional qubits [33]. Another widely used strategy involves classical preprocessing methods for high-dimensional data, such as principal component analysis (PCA) [34], to reduce input dimensionality before applying quantum encoding. This preprocessing step reduces overall quantum resource requirements while preserving relevant information.

Beyond fixed encoding strategies, learning-based approaches have emerged that dynamically adjust data encoding for specific tasks. For example, [35] achieve task-specific quantum embeddings by incorporating learnable parameters into the encoding layers, which are optimized to maximize class separability in Hilbert space. This technique is analogous to classical metric learning. Following this routine, a quantum few-shot embedding framework [36] has been proposed to encode classical data into quantum states, which can be generalized to the downstream quantum machine learning tasks. These methods enable quantum circuits to adapt their encodings dynamically, improving efficiency and performance.

2.6.2 Advanced Quantum Read-Out Methods

Conventional quantum read-out methods often face significant challenges, including high computational overhead and resource inefficiencies. Below, we discuss the primary challenges and discuss solutions in two quantum read-out methods: QST and observable estimation.

Challenge I: High computational overhead of QST. QST aims to reconstruct the density matrix of a quantum state, but this becomes computationally infeasible as the system size increases. This is because the required number of measurements and the classical memory grows exponentially with the number of qubits.

Challenge II: Resource inefficiency in observable estimation. The required number of measurements for observable estimation grows linearly with the number of Pauli terms in the observable. For observables where the number of Pauli terms substantially increases with the system size, the measurement cost becomes prohibitive.

For Challenge I, the key idea is to represent only a subspace of the quantum space. This effectively captures task-relevant properties while reducing computational cost. For example, in many QML algorithms, such as the HHL algorithm for solving linear systems [37] and quantum singular value decomposition [38], the solution state exists within the row or column space of the input matrix. When the input matrix is low-rank, state tomography can be obtained efficiently [39] as the linear combination of a complete basis chosen from the input matrix. Besides, an effective technique is matrix product state (MPS) tomography [40, 41]. This technique's key idea is that many practical quantum states, such as those in Ising models or low-entanglement systems, can be efficiently represented with fewer parameters. By focusing on states with limited entanglement, MPS tomography reconstructs the state using only a polynomial number of measurements with the qubit counts.

Another promising approach is using neural networks to parameterize quantum states. Neural quantum states allow for the efficient representation and reconstruction of density matrices, particularly for complex or high-dimensional quantum systems. For instance, restricted Boltzmann machines [42] and Transformer [43] have been applied to approximate the probability of measurement outcome and density matrices [44–47]. These approaches are particularly effective for systems that are difficult to capture using traditional methods.

For Challenge II, a measurement reduction technique can be applied by exploiting the commutativity of Pauli operators. When multiple Pauli terms commute, they can be measured simultaneously within the same measurement basis, significantly reducing the total measurement cost [48, 49]. This approach has been widely adopted in hybrid quantum-classical algorithms, such as variational quantum Eigensolvers (VQE) [50], where Hamiltonians are decomposed into sums of Pauli terms. Grouping commuting terms into clusters allows for efficient measurement strategies while preserving accuracy.

In addition to measurement grouping, adaptive measurement strategies further improve resource allocation during expectation value estimation. The key observation is that not all Pauli terms contribute equally to the total observable—terms with higher variance require more measurement shots for reliable estimation, while low-variance terms can be measured with fewer shots. Building on this insight, adaptive shot allocation techniques [51–53] dynamically distribute measurement resources across Pauli terms based on their statistical properties and achieve more accurate estimations with a finite measurement budget.

2.6.3 Advanced Quantum Linear Algebra

Quantum linear algebra, based on the block encoding and quantum singular value transformation framework, has proven its power for the design of quantum algorithms. Compared to the traditional subroutines like quantum phase estimation and quantum arithmetic [54, 55], quantum linear algebra can exponentially improve the dependency on precision [24]. However, a major drawback is that it can only deal with the singular values of block-encoded matrices.

A natural consideration is to generalize the singular value transformation to the eigenvalue transformation. One strong motivation from the application aspect for this is to solve the differential equations on the quantum computer [56–60]. This remains an active research field. Quantum eigenvalue processing, proposed by Low and Su [61], focuses on matrices with real spectra and Jordan forms, in which they prepare the Faber history state to achieve efficient eigenvalue transformation over the complex plane. An et al. [62] and [63] show that simulating a general class of non-unitary dynamics can be achieved by the linear combination of Hamiltonian simulation (LCHS).

Another approach is to broaden the range of functions that can be implemented by quantum linear algebra. Quantum phase processing, proposed by Wang et al. [64], can directly apply arbitrary trigonometric transformations to eigenphases of a unitary operator. Similar results have been independently obtained by Motlagh and Wiebe [65]. In addition, Rossi and Chuang [66] investigates how to implement multivariate functions. For the application, a representative example is the multivariate state preparation achieved by Mori et al. [67], enabling the amplitude encoding of classical multivariate data.

In Sect. 2.4, we introduce the concept of diagonal block encoding, which can convert a state preparation unitary into a block encoding. As the efficient construction of block encodings is a prerequisite for achieving end-to-end quantum advantage, an important research direction is to investigate which types of matrices can be efficiently prepared. By leveraging state-of-the-art techniques in quantum state preparation [68, 69] and the linear combination of unitaries [70], it is possible to efficiently construct block encodings for certain classes of matrices [71, 72]. Additionally, explicit constructions have been explored for specific types of sparse matrices [73].

References

1. Nielsen, M. A., & Chuang, I. L. (2011). *Quantum computation and quantum information: 10th anniversary edition* (2nd ed.). Cambridge University Press. ISBN 1-10-700217-6.
2. Arute, F., Arya, K., Babbush, R., Bacon, D., Bardin, J. C., Barends, R., Biswas, R., Boixo, S., Brandao, F. G. S. L., Buell, D. A., et al. (2019). Quantum supremacy using a programmable superconducting processor. *Nature, 574*(7779), 505–510.
3. Jozsa, R., & Linden, N. (2003). On the role of entanglement in quantum-computational speed-up. *Proceedings of the Royal Society of London. Series A: Mathematical, Physical and Engineering Sciences, 459*(2036), 2011–2032.
4. Greenberger, D. M., Horne, M. A., & Zeilinger, A. (1989). Going beyond bell's theorem. In *Bell's theorem, quantum theory and conceptions of the universe* (pp. 69–72). Springer.
5. Leach, D. P., & Malvino, A. P. (1994). *Digital principles and applications*. Glencoe/McGraw-Hill.
6. Dawson, C. M., & Nielsen, M. A. (2005). The solovay-kitaev algorithm. arXiv preprint quant-ph/0505030.
7. Wilde, M. M. (2011). From classical to quantum shannon theory. arXiv preprint arXiv:1106.1445.
8. Flammia, S. T., & Wallman, J. J. (2020). Efficient estimation of pauli channels. *ACM Transactions on Quantum Computing, 1*(1), 1–32.
9. Preskill, J. (1999). Lecture notes for physics 219: Quantum computation. *Caltech Lecture Notes, 7*, 1.
10. Aaronson, S. (2015). Read the fine print. *Nature Physics, 11*(4), 291–293.
11. Giovannetti, V., Lloyd, S., & Maccone, L. (2008). Quantum random access memory. *Physical Review Letters, 100*(16), 160501.
12. Vogel, K., & Risken, H. (1989). Determination of quasiprobability distributions in terms of probability distributions for the rotated quadrature phase. *Physical Review A, 40*(5), 2847.
13. Qi, B., Hou, Z., Li, L., Dong, D., Xiang, G., & Guo, G. (2013). Quantum state tomography via linear regression estimation. *Scientific Reports, 3*(1), 3496.
14. Hradil, Z. (1997). Quantum-state estimation. *Physical Review A, 55*(3), R1561.
15. Layden, D., Mazzola, G., Mishmash, R. V., Motta, M., Wocjan, P., Kim, J.-S., & Sheldon, S. (2023). Quantum-enhanced markov chain monte carlo. *Nature, 619*(7969), 282–287.
16. Farhi, E., Goldstone, J., & Gutmann, S. (2014). A quantum approximate optimization algorithm. arXiv preprint arXiv:1411.4028.
17. Grover, L. K. (1996). A fast quantum mechanical algorithm for database search. In *Proceedings of the Twenty-Eighth Annual ACM Symposium on Theory of Computing* (pp. 212–219).
18. Boixo, S., Isakov, S. V., Smelyanskiy, V. N., Babbush, R., Ding, N., Jiang, Z., Bremner, M. J., Martinis, J. M., & Neven, H. (2018). Characterizing quantum supremacy in near-term devices. *Nature Physics, 14*(6), 595–600.
19. Bouland, A., Fefferman, B., Nirkhe, C., & Vazirani, U. (2019). On the complexity and verification of quantum random circuit sampling. *Nature Physics, 15*(2), 159–163.
20. Kandala, A., Mezzacapo, A., Temme, K., Takita, M., Brink, M., Chow, J. M., & Gambetta, J. M. (2017). Hardware-efficient variational quantum eigensolver for small molecules and quantum magnets. *Nature, 549*(7671), 242–246.
21. Tilly, J., Chen, H., Cao, S., Picozzi, D., Setia, K., Li, Y., Grant, E., Wossnig, L., Rungger, I., Booth, G. H., et al. (2022). The variational quantum eigensolver: A review of methods and best practices. *Physics Reports, 986*, 1–128.
22. Aaronson, S. (2018). Shadow tomography of quantum states. In *Proceedings of the 50th Annual ACM SIGACT Symposium on Theory of Computing* (pp. 325–338).
23. Huang, H.-Y., Kueng, R., & Preskill, J. (2020). Predicting many properties of a quantum system from very few measurements. *Nature Physics, 16*(10), 1050–1057.

24. Gilyén, A., Su, Y., Low, G. H., & Wiebe, N. (2019). Quantum singular value transformation and beyond: Exponential improvements for quantum matrix arithmetics. In *Proceedings of the 51st Annual ACM SIGACT Symposium on Theory of Computing, STOC '19*. ACM. https://doi.org/10.1145/3313276.3316366

25. Guo, N., Yu, Z., Choi, M., Agrawal, A., Nakaji, K., Aspuru-Guzik, A., & Rebentrost, P. (2024). Quantum linear algebra is all you need for transformer architectures. https://arxiv.org/abs/2402.16714

26. Brassard, G., Høyer, P., Mosca, M., & Tapp, A. (2002). Quantum amplitude amplification and estimation. In *Quantum computation and information (Washington, DC, 2000)* (Vol. 305). Contemporary Mathematics (pp. 53–74). American Mathematical Society. ISBN 978-0-8218-2140-4. https://doi.org/10.1090/conm/305/05215

27. Zhao, L., Zhao, Z., Rebentrost, P., & Fitzsimons, J. (2021). Compiling basic linear algebra subroutines for quantum computers. *Quantum Machine Intelligence, 3*(2), 21. ISSN 2524-4914. https://doi.org/10.1007/s42484-021-00048-8

28. Guo, N., Mitarai, K., & Fujii, K. (2024). Nonlinear transformation of complex amplitudes via quantum singular value transformation. *Physical Review Research, 6*, 043227. https://doi.org/10.1103/PhysRevResearch.6.043227

29. Rattew, A. G., & Rebentrost, P. (2023). Non-linear transformations of quantum amplitudes: Exponential improvement, generalization, and applications. https://arxiv.org/abs/2309.09839

30. Nakaji, K., Uno, S., Suzuki, Y., Raymond, R., Onodera, T., Tanaka, T., Tezuka, H., Mitsuda, N., & Yamamoto, N. (2022). Approximate amplitude encoding in shallow parameterized quantum circuits and its application to financial market indicators. *Physical Review Research, 4*(2), 023136.

31. Pérez-Salinas, A., Cervera-Lierta, A., Gil-Fuster, E., & Latorre, J. I. (2020). Data re-uploading for a universal quantum classifier. *Quantum, 4*, 226.

32. Hur, T., Araujo, I. F., & Park, D. K. (2024). Neural quantum embedding: Pushing the limits of quantum supervised learning. *Physical Review A, 110*(2), 022411.

33. Schuld, M., & Petruccione, F. (2018). Information encoding. In M. Schuld & F. Petruccione (Eds.), *Supervised learning with quantum computers* (pp. 139–171). Springer.

34. Abdi, H., & Williams, L. J. (2010). Principal component analysis. *Wiley Interdisciplinary Reviews: Computational Statistics, 2*(4), 433–459.

35. Lloyd, S., Schuld, M., Ijaz, A., Izaac, J., & Killoran, N. (2020). Quantum embeddings for machine learning. arXiv preprint arXiv:2001.03622.

36. Liu, M., Liu, J., Liu, R., Makhanov, H., Lykov, D., Apte, A., & Alexeev, Y. (2022). Embedding learning in hybrid quantum-classical neural networks. In *2022 IEEE International Conference on Quantum Computing and Engineering (QCE)* (pp. 79–86). IEEE.

37. Harrow, A. W., Hassidim, A., & Lloyd, S. (2009). Quantum algorithm for linear systems of equations. *Physical Review Letters, 103*(15), 150502.

38. Rebentrost, P., Steffens, A., Marvian, I., & Lloyd, S. (2018). Quantum singular-value decomposition of nonsparse low-rank matrices. *Physical Review A, 97*(1), 012327.

39. Zhang, K., Hsieh, M.-H., Liu, L., & Tao, D. (2021). Quantum gram-schmidt processes and their application to efficient state readout for quantum algorithms. *Physical Review Research, 3*(4), 043095.

40. Lanyon, B. P., Maier, C., Holzäpfel, M., Baumgratz, T., Hempel, C., Jurcevic, P., Dhand, I., Buyskikh, A.S., Daley, A. J., ramer, M., et al. (2017). Efficient tomography of a quantum many-body system. *Nature Physics, 13*(12), 1158–1162.

41. Orús, R. (2019). Tensor networks for complex quantum systems. *Nature Reviews Physics, 1*(9), 538–550.

42. Fischer, A., & Igel, C. (2012). An introduction to restricted boltzmann machines. In *Progress in Pattern Recognition, Image Analysis, Computer Vision, and Applications: 17th Iberoamerican Congress, CIARP 2012*, Buenos Aires, Argentina, September 3–6, 2012. Proceedings 17 (pp. 14–36). Springer.

43. Vaswani, A. (2017). Attention is all you need. In *Advances in neural information processing systems*.

44. Torlai, G., Mazzola, G., Carrasquilla, J., Troyer, M., Melko, R., & Carleo, G. (2018). Neural-network quantum state tomography. *Nature Physics, 14*(5), 447–450. https://doi.org/10.1038/s41567-018-0048-5

45. Schmale, T., Reh, M., & Gärttner, M. (2022). Efficient quantum state tomography with convolutional neural networks. *NPJ Quantum Information, 8*(1). ISSN 2056-6387. https://doi.org/10.1038/s41534-022-00621-4

46. Wang, H., Weber, M., Izaac, J., & Lin, C. Y.-Y. (2022). Predicting properties of quantum systems with conditional generative models. arXiv preprint arXiv:2211.16943.

47. Zhao, L., Guo, N., Luo, M.-X., & Rebentrost, P. (2023). Provable learning of quantum states with graphical models. https://arxiv.org/abs/2309.09235

48. Kandala, A., Mezzacapo, A., Temme, K., Takita, M., Brink, M., Chow, J. M., & Gambetta, J. M. (2017). Hardware-efficient variational quantum eigensolver for small molecules and quantum magnets. *Nature, 549*(7671), 242–246. ISSN 0028-0836, 1476-4687. https://doi.org/10.1038/nature23879

49. Verteletskyi, V., Yen, T.-C., & Izmaylov, A. F. (2020). Measurement optimization in the variational quantum eigensolver using a minimum clique cover. *The Journal of Chemical Physics, 152*(12), 124114.

50. Cerezo, M., Arrasmith, A., Babbush, R., Benjamin, S. C., Endo, S., Fujii, K., McClean, J. R., Mitarai, K., Yuan, X., Cincio, L., et al. (2021). Variational quantum algorithms. *Nature Reviews Physics, 3*(9), 625–644.

51. Rubin, N. C., Babbush, R., & McClean, J. (2018). Application of fermionic marginal constraints to hybrid quantum algorithms. *New Journal of Physics, 20*(5), 053020.

52. Arrasmith, A., Cincio, L., Somma, R. D., & Coles, P. J. (2020). Operator sampling for shot-frugal optimization in variational algorithms. arXiv preprint arXiv:2004.06252.

53. Qian, Y., Du, Y., & Tao, D. (2024). Shuffle-qudio: Accelerate distributed vqe with trainability enhancement and measurement reduction. *Quantum Machine Intelligence, 6*(1), 1–22.

54. Kitaev, A. Y. (1995). Quantum measurements and the abelian stabilizer problem. https://arxiv.org/abs/quant-ph/9511026

55. Ruiz-Perez, L., & Garcia-Escartin, J. C. (2017). Quantum arithmetic with the quantum fourier transform. *Quantum Information Processing, 16*(6). ISSN 1573-1332. https://doi.org/10.1007/s11128-017-1603-1

56. Liu, J.-P. Kolden, H. Ø., Krovi, H. K., Loureiro, N. F., Trivisa, K., & Childs, A. M. (2021). Efficient quantum algorithm for dissipative nonlinear differential equations. *Proceedings of the National Academy of Sciences, 118*(35). ISSN 1091-6490. https://doi.org/10.1073/pnas.2026805118

57. Childs, A. M., Liu, J.-P., & Ostrander, A. (2021). High-precision quantum algorithms for partial differential equations. *Quantum, 5*, 574. ISSN 2521-327X. https://doi.org/10.22331/q-2021-11-10-574

58. An, D., Linden, N., Liu, J.-P., Montanaro, A., Shao, C., & Wang, J. (2021). Quantum-accelerated multilevel Monte Carlo methods for stochastic differential equations in mathematical finance. *Quantum, 5*, 481. ISSN 2521-327X. https://doi.org/10.22331/q-2021-06-24-481

59. Jin, S., Liu, N., & Yu, Y. (2022). Quantum simulation of partial differential equations via schrodingerisation. https://arxiv.org/abs/2212.13969

60. Shang, Z.-X., Guo, N., An, D., & Zhao, Q. (2024). Design nearly optimal quantum algorithm for linear differential equations via lindbladians. https://arxiv.org/abs/2410.19628

61. Low, G. H., & Su, Y. (2024). Quantum eigenvalue processing. https://arxiv.org/abs/2401.06240

62. An, D., Liu, J.-P., & Lin, L. (2023). Linear combination of Hamiltonian simulation for nonunitary dynamics with optimal state preparation cost. *Physical Review Letters, 131*, 150603. https://doi.org/10.1103/PhysRevLett.131.150603

63. An, D., Childs, A. M., Lin, L., & Ying, L. (2024). Laplace transform based quantum eigenvalue transformation via linear combination of Hamiltonian simulation. https://arxiv.org/abs/2411.04010

64. Wang, Y., Zhang, L., Yu, Z., & Wang, X. (2023). Quantum phase processing and its applications in estimating phase and entropies. *Physical Review A, 108*(6). ISSN 2469-9934. https://doi.org/10.1103/PhysRevA.108.062413
65. Motlagh, D., & Wiebe, N. (2024). Generalized quantum signal processing. https://arxiv.org/abs/2308.01501
66. Rossi, Z. M., & Chuang, I. L. (2022). Multivariable quantum signal processing (m-QSP): Prophecies of the two-headed oracle. *Quantum, 6*, 811. ISSN 2521-327X. https://doi.org/10.22331/q-2022-09-20-811
67. Mori, H., Mitarai, K., & Fujii, K. (2024). Efficient state preparation for multivariate monte carlo simulation. https://arxiv.org/abs/2409.07336
68. Zhang, X-M., Li, T., & Yuan, X. (2022). Quantum state preparation with optimal circuit depth: Implementations and applications. *Physical Review Letters, 129*(23), 230504. https://doi.org/10.1103/PhysRevLett.129.230504
69. Sun, X., Tian, G., Yang, S., Yuan, P., & Zhang, S. (2023). Asymptotically optimal circuit depth for quantum state preparation and general unitary synthesis. *IEEE Transactions on Computer-Aided Design of Integrated Circuits and Systems, 42*(10), 3301–3314. ISSN 1937-4151. https://doi.org/10.1109/TCAD.2023.3244885. https://ieeexplore.ieee.org/document/10044235
70. Childs, A. M., & Wiebe, N. (2012). Hamiltonian simulation using linear combinations of unitary operations. *Quantum Information & Computation, 12*(11–12), 901–924. ISSN 1533-7146.
71. Guseynov, N., & Liu, N. (2024). Efficient explicit circuit for quantum state preparation of piece-wise continuous functions. https://arxiv.org/abs/2411.01131
72. Guseynov, N., Huang, X., & Liu, N. (2024). Explicit gate construction of block encoding for Hamiltonians needed for simulating partial differential equations. https://arxiv.org/abs/2405.12855
73. Camps, D., Lin, L., Van Beeumen, R., & Yang, C. (2023). Explicit quantum circuits for block encodings of certain sparse matrices. https://arxiv.org/abs/2203.10236

Chapter 3
Quantum Kernel Methods

Abstract This chapter provides a comprehensive guide to understanding quantum kernel methods, covering the fundamental concepts of classical and quantum kernel methods, their theoretical foundations, and practical implementations. Section 3.1 offers a detailed introduction to classical kernel methods, including their motivation, derivation, and the construction of classical kernel functions. Building on this foundation, Sect. 3.2 discusses the motivation for implementing kernel methods on quantum devices, exploring the potential advantages of quantum kernels. It also introduces the specific implementation of quantum kernel functions, clarifies the relationship between classical and quantum kernel machines, and provides concrete examples of quantum kernels. Section 3.3 delves into the theoretical foundations of quantum kernels, focusing on two key aspects: the expressivity and generalization properties of quantum kernel machines. It examines the diverse feature spaces that quantum kernels can represent and the potential advantages of quantum kernels in reducing generalization error compared to classical kernel methods. This analysis underscores the ability of quantum kernels to improve the accuracy of predictions for unseen data. Finally, Sect. 3.4 demonstrates simple yet illustrative code implementations of quantum kernels using the MNIST dataset.

Machine learning (ML) algorithms fundamentally aim to learn underlying feature representations within training data. This allows data points to be effectively modeled using simple tools like linear classifiers. **Kernel methods** offer a powerful way to achieve this. They capture nonlinear patterns efficiently. In kernel methods, a kernel function is defined as the inner product between the high-dimensional feature representations of data points. These feature representations are generated by a hidden feature map, which transforms the original data into a higher-dimensional space. In this space, complex patterns become easier to identify and model. The kernel function, therefore, serves as a measure of similarity between data points in this transformed space.

The effectiveness of kernel methods heavily depends on the hidden feature map's ability to capture relevant patterns in the data. The better this feature map reveals the underlying structure, the better the kernel method performs in learning and

© The Author(s), under exclusive license to Springer Nature Singapore Pte Ltd. 2025
Y. Du et al., *A Gentle Introduction to Quantum Machine Learning*,
https://doi.org/10.1007/978-981-95-1284-3_3

generalizing from data. However, classical kernel methods are inherently limited by the types of patterns they can recognize, as these are constrained by classical computational frameworks. Essentially, classical models excel at detecting patterns they are specifically designed to recognize but may struggle with patterns that deviate from this framework.

In contrast, quantum mechanics can generate complex, nonintuitive patterns often beyond classical algorithms' reach. Quantum systems can produce statistical correlations that are computationally challenging—or even impossible—for classical computers to replicate. This suggests that employing quantum circuits as hidden feature maps could enable the detection of patterns that are difficult or impractical for classical models to capture. By leveraging quantum circuits, we can potentially access new regions of the feature space, leading to improved pattern recognition capabilities and, consequently, better learning performance.

These insights motivate the development of **quantum kernel methods**, where both the hidden feature map and the kernel function are implemented on a quantum computer. By harnessing the unique properties of quantum mechanics, such as superposition and entanglement, quantum kernel methods have the potential to surpass their classical counterparts in specific machine learning tasks. This could result in more powerful models with enhanced generalization capabilities.

In this chapter, we provide a step-by-step explanation of the transition from classical kernel machines to quantum kernel machines in Sects. 3.1 and 3.2. Moreover, we discuss the theoretical foundation of quantum kernel machines in Sect. 3.3 from the aspects of expressivity and generalization of quantum kernel machines. Finally, we demonstrate simple yet illustrative code implementations on MNIST dataset.

3.1　Classical Kernel Machines

3.1.1　Motivation of Kernel Methods

Before delving into kernel machines, it is essential to first understand the motivation behind kernel methods. In many machine learning tasks, particularly in classification, the goal is to find a decision boundary that best separates different classes of data. When the data is linearly separable, this boundary can be represented as a straight line (in 2D), a plane (in 3D), or a hyperplane (in higher dimensions), as illustrated in Fig. 3.1a. Mathematically, given an input space $\mathcal{X} \subset \mathbb{R}^d$ with $d \geq 1$ and a target or output space $\mathcal{Y} = \{+1, -1\}$, we consider a training dataset $\mathcal{D} = \{(x^{(i)}, y^{(i)})\}_{i=1}^{n} \in (\mathcal{X} \times \mathcal{Y})^n$ where each data point $x^{(i)} \in \mathcal{X}$ is associated with a label $y^{(i)} \in \mathcal{Y}$. For the dataset to be linearly separable, there must exist a vector $w \in \mathbb{R}^d$ and a bias term $b \in \mathbb{R}$ such that

$$\forall i \in [n], \quad y^{(i)}(w^{\top}x^{(i)} + b) \geq 0, \tag{3.1}$$

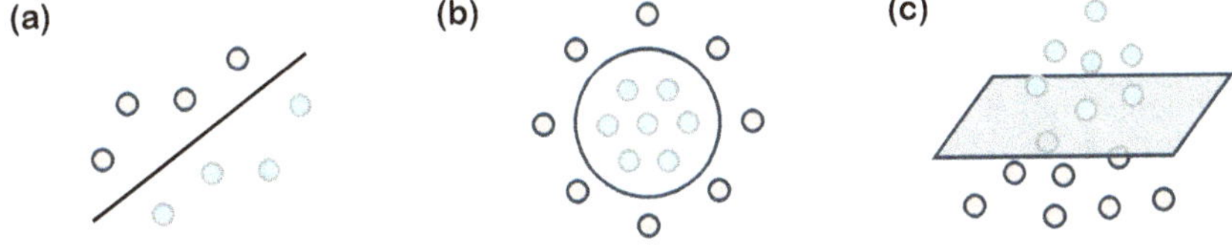

Fig. 3.1 Various distributions of data points

where $\boldsymbol{w}^\top \boldsymbol{x}^{(i)}$ represents the inner product of vectors $\boldsymbol{w}$ and $\boldsymbol{x}^{(i)}$. This means that a hyperplane defined by $(\boldsymbol{w}, b)$ can perfectly separate the two classes.

However, in real-world scenarios, data is often not linearly separable, as shown in Fig. 3.1b. The decision boundary required to separate classes may be curved or highly complex. Traditional linear models struggle with such nonlinear data because they are inherently limited to creating only linear decision boundaries. This limitation highlights the need for more flexible approaches.

To address nonlinear data, one effective strategy is to transform the input data into a higher-dimensional space where it might become linearly separable. This transformation is known as *feature mapping*, denoted by

$$\phi : \boldsymbol{x} \rightarrow \phi(\boldsymbol{x}) \in \mathbb{R}^D. \tag{3.2}$$

Here, the original input space $\mathcal{X}$ is mapped to a higher-dimensional feature space $\mathbb{R}^D$, where $D \geq d$. The idea is that in this higher-dimensional space, linear models can more easily identify complex patterns from the original data. A visualization is shown in as shown in Fig. 3.1c.

However, explicitly computing the feature map $\phi(\boldsymbol{x})$ in Eq. (3.2) can be computationally expensive, especially if the feature space is high dimensional or even infinite dimensional. Fortunately, many machine learning algorithms for tasks like classification or regression depend primarily on the inner product between data points, which will be explained in Sect. 3.1.2. In the feature space, this inner product is given by $\langle \phi(\boldsymbol{x}^{(i)}), \phi(\boldsymbol{x}^{(j)}) \rangle$.

Remark
Throughout the whole tutorial, we interchangeably use $\boldsymbol{a}^\top \boldsymbol{b}$, $\boldsymbol{a} \cdot \boldsymbol{b}$, $\langle \boldsymbol{a}, \boldsymbol{b} \rangle$, and $\langle \boldsymbol{a} | \boldsymbol{b} \rangle$ to denote the inner production of two vectors $\boldsymbol{a}$ and $\boldsymbol{b}$.

To circumvent the computational cost of explicitly calculating the feature map, we can use a *kernel function*. A kernel function $k(\boldsymbol{x}^{(i)}, \boldsymbol{x}^{(j)})$ is defined as

$$k(\boldsymbol{x}^{(i)}, \boldsymbol{x}^{(j)}) = \langle \phi(\boldsymbol{x}^{(i)}), \phi(\boldsymbol{x}^{(j)}) \rangle. \tag{3.3}$$

This allows us to compute the inner product in the higher-dimensional feature space indirectly, without ever having to compute $\phi(x)$ explicitly. This approach is commonly known as the *kernel trick*.

Using the kernel function directly in algorithms avoids the computational overhead of working in a high-dimensional space. The collection of kernel values for a dataset forms the kernel matrix (or Gram matrix), where each entry is given by $k(x^{(i)}, x^{(j)})$. This matrix is central to many kernel-based algorithms, as it captures the pairwise similarities between all training data points.

To illustrate the kernel trick, let's consider a simple example in a d-dimensional input space, where $x = (x_1, \cdots, x_d)^\top$. Suppose the polynomial kernel is of degree 2, defined as

$$
\begin{aligned}
k(x, z) &= (x^\top z)^2 \\
&= (x_1 z_1 + \cdots + x_d z_d)^2 \\
&= \sum_{i=1}^{d} \sum_{j=1}^{d} x_i z_i x_j z_j \\
&= [x_1^2, \cdots, x_d^2, \sqrt{2} x_1 x_2, \cdots, \sqrt{2} x_d x_{d-1}] \\
&\quad\; [z_1^2, \cdots, z_d^2, \sqrt{2} z_1 z_2, \cdots, \sqrt{2} z_d z_{d-1}]^\top \\
&= \phi(x)^\top \phi(z).
\end{aligned}
\tag{3.4}
$$

Here, it can be seen that the feature mapping, which comprises all second-order terms, takes the form as

$$
\phi(x) = [x_1^2, \cdots, x_d^2, \sqrt{2} x_1 x_2, \cdots, \sqrt{2} x_d x_{d-1}]^\top.
\tag{3.5}
$$

Notably, directly computing the kernel function $(x^\top z)^2$ for a large d is much more efficient than explicitly calculating the feature map $\phi(x)$ and then taking the inner product $\phi(x)^\top \phi(z)$. Specifically, using the kernel function only requires $O(d)$ time, since it involves computing the dot product in the original input space $\mathbb{R}^d$. In contrast, explicitly computing the transformed feature vectors $\phi(x)$ and their inner product could increase time complexity to $O(D)$, where D is the dimensionality of the feature space after mapping. For this example of a polynomial kernel with degree 2, D can grow to $O(d^2)$. This demonstrates the kernel trick's computational efficiency.

Remark
Throughout this manuscript, we use the notations O and Ω to represent the asymptotic upper and lower bounds, respectively, on the growth rate of a term, ignoring constant factors and lower-order terms.

3.1.2 Dual Representation

To understand why many machine learning algorithms rely primarily on the inner products between data points, we need to introduce the concept of the *dual representation*. In essence, many linear parametric models used for regression or classification can be recast into an equivalent dual form, where the kernel function evaluated on the training data emerges naturally.

Let's start with a linear regression model with the training dataset $\{(x^{(i)}, y^{(i)})\}_{i=1}^{n}$. Here, the parameters are determined by minimizing a regularized sum-of-squares error function:

$$\mathcal{L}(w) = \frac{1}{2} \sum_{i=1}^{n} \left(w^{\top} \phi(x^{(i)}) - y^{(i)} \right)^2 + \frac{\lambda}{2} w^{\top} w, \tag{3.6}$$

where $w^{\top}$ refers to the transpose of model parameters w, $\phi(x^{(i)})$ represents the feature mapping of the input $x^{(i)}$, and $\lambda \geq 0$ is the regularization factor that helps prevent overfitting.

To find the optimal w, the gradient of $\mathcal{L}(w)$ with respect to w is set to be zero, i.e.,

$$\frac{\partial \mathcal{L}(w)}{\partial w} = \sum_{i=1}^{n} \left(w^{\top} \phi(x^{(i)}) - y^{(i)} \right) \phi(x^{(i)}) + \lambda w = 0. \tag{3.7}$$

From this, it can be seen that the solution for w can be expressed as a linear combination of the training data's feature vectors:

$$w = -\frac{1}{\lambda} \sum_{i=1}^{n} (w^{\top} \phi(x^{(i)}) - y^{(i)}) \phi(x^{(i)}) = \sum_{i=1}^{n} a^{(i)} \phi(x^{(i)}) := \Phi^{\top} a, \tag{3.8}$$

where $\Phi = \left[\phi(x^{(1)}), \cdots, \phi(x^{(n)}) \right]^{\top}$ is the design matrix, whose i-th row is given by $\phi(x^{(i)})^{\top}$. Here, the coefficients $a^{(i)}$ are functions of w, defined as

$$a^{(i)} = -\frac{1}{\lambda} \left(w^{\top} \phi(x^{(i)}) - y^{(i)} \right). \tag{3.9}$$

Thus, instead of directly optimizing w, the problem can be reformulated in terms of the parameter vector a, giving rise to a dual representation. By substituting $w = \Phi^\top a$ into the original objective function $\mathcal{L}(w)$ in Eq. (3.6), the result is

$$\mathcal{L}(a) = \frac{1}{2}a^\top \Phi\Phi^\top\Phi\Phi^\top a - a^\top\Phi\Phi^\top y + \frac{1}{2}y^\top y + \frac{\lambda}{2}a^\top\Phi\Phi^\top y, \tag{3.10}$$

where $y = \left(y^{(1)}, \cdots, y^{(n)}\right)^\top$ denotes the vector representation of n training labels. The kernel matrix is defined by $K = \Phi\Phi^\top$, where each element is given by

$$K_{ij} = \phi(x^{(i)})^\top\phi(x^{(j)}) = k(x^{(i)}, x^{(j)}), \tag{3.11}$$

using kernel function $k(x, x')$ defined by Eq. (3.3). The objective function in terms of a simplifies to

$$\mathcal{L}(a) = \frac{1}{2}a^\top K^2 a - a^\top K y + \frac{1}{2}y^\top y + \frac{\lambda}{2}a^\top K y. \tag{3.12}$$

Setting the gradient of $\mathcal{L}(a)$ with respect to a to zero gives us

$$a = (K + \lambda\mathbb{I}_n)^{-1}y, \tag{3.13}$$

where $\mathbb{I}_n$ is the identity matrix of size $n \times n$.

Now, using this dual formulation, it can be derived the prediction for a new input x. Substituting $w = \Phi^\top a$ in Eq. (3.8), the prediction of x is given by

$$y(x) = w^\top\phi(x) = \langle\Phi^\top a, \phi(x)\rangle = k(x)^\top(K + \lambda\mathbb{I}_n)^{-1}y, \tag{3.14}$$

where $k(x) \in \mathbb{R}^n$ is a vector with elements $k_i(x) = k(x^{(i)}, x) = \phi(x^{(i)})^\top\phi(x)$. This shows that the dual formulation allows us to express the solution entirely in terms of the kernel function $k(x, x')$, rather than explicitly working with the feature map $\phi(x)$. This is particularly advantageous because it enables us to work in high-dimensional or even infinite-dimensional feature spaces implicitly.

In the dual formulation, the parameter vector a is determined by inverting an $n \times n$ matrix. The original parameter space formulation, in contrast, requires inverting a $d \times d$ matrix to determine w. Although this may not seem advantageous when $n > d$, the true benefit of the dual formulation lies in its ability to leverage the kernel trick. By expressing the solution in terms of the kernel function, we avoid the explicit computation of the feature vectors $\phi(x)$. This allows us to implicitly utilize feature spaces of very high, or even infinite, dimensionality, enabling the model to capture complex, nonlinear relationships in the data without the associated computational cost.

> **Remark**
> We standardize the notation used throughout this chapter to help readers
> follow the content more easily. The kernel function is represented by the
> lowercase letter k or with subscripts k_Q and k_C. The kernel matrix is denoted
> by the capital letter K or with subscripts K_Q and K_C. Additionally, we use
> the bold lowercase letter $\boldsymbol{k}(\boldsymbol{x})$ to represent the vector of kernel values, where
> each element is given by $\boldsymbol{k}_j(\boldsymbol{x}) = k(\boldsymbol{x}^{(j)}, \boldsymbol{x})$, corresponding to the training
> points $\boldsymbol{x}^{(j)} \in \{\boldsymbol{x}^{(1)}, \ldots, \boldsymbol{x}^{(n)}\}$.

3.1.3   Kernel Construction

To utilize the kernel trick in machine learning algorithms, it is essential to construct
valid kernel functions. One approach is to start with a feature mapping $\phi(\boldsymbol{x})$ and
then derive the corresponding kernel. For a one-dimensional input space, the kernel
function is defined as

$$k(\boldsymbol{x}, \boldsymbol{x}') = \phi(\boldsymbol{x})^\top \phi(\boldsymbol{x}') = \sum_{i=1}^{D} \langle \phi_i(\boldsymbol{x}), \phi_i(\boldsymbol{x}') \rangle, \tag{3.15}$$

where $\phi_i(\boldsymbol{x})$ are the basis functions of the feature map.

Alternatively, kernels can be constructed directly without explicitly defining a
feature map. In this case, the chosen function must be a valid kernel. This means
it needs to correspond to an inner product in some (possibly infinite-dimensional)
feature space. Mercer's condition guarantees a kernel function's validity.

Fact 3.1 (Mercer's Condition) *Let $\mathcal{X} \subset \mathbb{R}^d$ be a compact set and let $k : \mathcal{X} \times \mathcal{X} \to$*
$\mathbb{R}$ be a continuous and symmetric function. Then, k admits a uniformly convergent
expansion of the form

$$k(\boldsymbol{x}, \boldsymbol{x}') = \sum_{i=0}^{\infty} a_i \langle \phi_i(\boldsymbol{x}), \phi_i(\boldsymbol{x}') \rangle \tag{3.16}$$

with $a_i > 0$ if and only if for any square-integrable function c, the following
condition holds

$$\int_{\mathcal{X}} \int_{\mathcal{X}} c(\boldsymbol{x}) c(\boldsymbol{x}') k(\boldsymbol{x}, \boldsymbol{x}') \mathrm{d}\boldsymbol{x} \mathrm{d}\boldsymbol{x}' \geq 0. \tag{3.17}$$

Mercer's condition is crucial because it ensures that the optimization problem for
algorithms like support vector machines (SVM) remains convex [1], guaranteeing

convergence to a global minimum. A condition equivalent to Mercer's condition (under the assumptions of the theorem) is that the kernel $k(\cdot, \cdot)$ be positive definite symmetric. This property is more general since it does not require any assumption about $\mathcal{X}$.

Definition 3.1 (Positive Definite Symmetric Kernels) A kernel $k : \mathcal{X} \times \mathcal{X} \to \mathbb{R}$ is said to be positive definite symmetric (PDS) if for any $\{x^{(1)}, \cdots, x^{(n)}\} \subset \mathcal{X}$, the matrix $K = [k(x^{(i)}, x^{(j)})]_{ij} \in \mathbb{R}^{n \times n}$ is symmetric positive semi-definite (SPSD).

In other words, a kernel matrix K associated with a PDS kernel function will always be SPSD, ensuring that the corresponding optimization problem remains well behaved.

Below, several commonly used positive definite symmetric kernels are presented.

Example 3.1 (Polynomial Kernels) A polynomial kernel of degree $m \in \mathbb{N}$ with a constant $c > 0$ is defined as

$$k(x, x') = (x \cdot x' + c)^m, \quad \forall x, x' \in \mathbb{R}^d. \tag{3.18}$$

This kernel maps the input space to a higher-dimensional space of dimension $\binom{d+m}{m}$. For instance, in a two-dimensional input space ($d = 2$) and with $m = 2$, the kernel is expanded as follows:

$$k(x, x') = (x_1 x_1' + x_2 x_2' + c)^2 = \begin{bmatrix} x_1^2 \\ x_2^2 \\ \sqrt{2}\, x_1 x_2 \\ \sqrt{2c}\, x_1 \\ \sqrt{2c}\, x_2 \\ c \end{bmatrix} \cdot \begin{bmatrix} x_1'^2 \\ x_2'^2 \\ \sqrt{2}\, x_1' x_2' \\ \sqrt{2c}\, x_1' \\ \sqrt{2c}\, x_2' \\ c \end{bmatrix}. \tag{3.19}$$

In other words, this kernel corresponds to an inner product in a higher-dimensional space of dimension 6.

Thus, the features corresponding to a second-degree polynomial include the original features (x_1 and x_2), products of these features, and the constant feature. More generally, the features associated with a polynomial kernel of degree m are all the monomials of degree at most m based on the original features.

Example 3.2 (Gaussian Kernels) The Gaussian kernel (or Radial Basis Function, RBF) is one of the most widely used kernels, defined as

$$\forall x, x' \in \mathbb{R}^d, \quad k(x, x') = \exp\left(-\frac{\|x - x'\|}{2\sigma^2}\right), \tag{3.20}$$

where $\sigma > 0$ controls the width of the Gaussian function.

Gaussian kernels are particularly effective in capturing complex nonlinear patterns due to their infinite-dimensional feature space.

Example 3.3 (Sigmoid Kernels) The sigmoid kernel over $\mathbb{R}^d$ is defined as

$$k(x, x') = \tanh(a \cdot (x \cdot x') + b), \quad \forall x, x' \in \mathbb{R}^d, \tag{3.21}$$

where $a, b > 0$ are constants and $\tanh(c) = \frac{e^c - e^{-c}}{e^c + e^{-c}}$ is the hyperbolic tangent function, which squashes an arbitrary constant $c \in \mathbb{R}$ to a value between -1 and 1.

This kernel relates to neural networks, as it resembles the activation function commonly used in multilayer perceptrons, as introduced in the next chapter. Using the sigmoid kernel with support vector machines results in a model similar to a simple neural network.

Remark
Support vector machines (SVMs) are a well-known algorithm that heavily relies on kernel methods and are primarily used for classification tasks. The objective of an SVM is to identify the optimal hyperplane that separates data points from different classes with the *maximum margin*. The margin is defined as the distance between the hyperplane and the closest data points from each class, known as support vectors. SVMs can be applied to both linear and nonlinear classification problems. For nonlinear cases, kernel functions are employed to map the data into higher-dimensional spaces, enabling the separation of data that is not linearly separable in the original space. For a detailed introduction to SVMs, please refer to [1].

3.2 Quantum Kernel Machines

Quantum machine learning might someday outperform classical ML, but researchers need to tackle big hurdles first. To effectively introduce quantum kernel machines, it is essential to recognize the limitations of classical kernel machines. As discussed in Sect. 3.1, classical kernel machines rely on manually tailored feature mappings, such as polynomials or radial basis functions. However, these mappings may fail to capture the complex patterns within the dataset. Quantum kernel machines emerge as a promising alternative, as they perform feature mapping using quantum circuits, enabling them to explore exponentially larger feature spaces that are otherwise infeasible for classical computation.

Another crucial characteristic of quantum kernels is that they can be effectively implemented on near-term quantum devices, making them a practical tool for exploring the utility of near-term quantum technologies.

3.2.1 Quantum Feature Maps and Quantum Kernel Machines

The key difference between quantum kernel machines and classical kernel machines lies in how the feature mapping is performed. In the quantum context, a feature map refers to the injective encoding of classical data $x \in \mathbb{R}^d$ into a quantum state $|\phi(x)\rangle = U(x)|\psi\rangle$ on an N-qubit quantum register, where $U(x)$ refers to the physical operation or quantum circuit that depends on the data x. This feature map is implemented on a quantum computer and produces quantum states, which are referred to as quantum feature maps.

Definition 3.2 (Quantum feature map) Given an N-qubit quantum system initialized in state $|\psi\rangle$, let $x \in \mathcal{X} \subset \mathbb{R}^d$ be classical data. The quantum feature map is defined as the mapping

$$\phi : \mathcal{X} \to \mathcal{F},$$

$$\phi(x) = |\phi(x)\rangle \langle \phi(x)| = \rho(x), \tag{3.22}$$

where $\mathcal{F}$ is the space of complex-valued $2^N \times 2^N$ matrices equipped with the Hilbert-Schmidt inner product $\langle \rho, \sigma \rangle = \text{Tr}(\rho\sigma)$ for $\rho, \sigma \in \mathcal{F}$. In addition, the state $|\phi(x)\rangle$ can be implemented by applying a data-encoding quantum circuit $U(x)$ introduced in Sect. 2.3.1 on an initial state $|\psi\rangle$, leading to the expression of $|\phi(x)\rangle = U(x)|\psi\rangle$.

Recall that one way of constructing kernels is adopting the inner product of the defined feature mappings. Using the Hilbert-Schmidt inner product from Definition 3.2, the quantum kernel is defined as follows.

Definition 3.3 (Quantum kernel) Let ϕ be a quantum feature map over the domain $\mathcal{X}$. The quantum kernel k_Q is the inner product between two quantum feature maps $\rho(x)$ and $\rho(x')$ for data points $x, x' \in \mathcal{X}$:

$$k_Q : \mathcal{X} \times \mathcal{X} \to \mathbb{R},$$

$$k_Q(x, x') = \mathrm{Tr}(\rho(x)\rho(x')) = \left|\langle \phi(x)|\phi(x')\rangle\right|^2. \tag{3.23}$$

To justify the term "kernel," the quantum kernel must be shown to be a *positive definite function*. A quantum kernel can be expressed as the product of a complex-valued kernel $\hat{k}_Q(x, x') = \langle \phi(x)|\phi(x')\rangle \in \mathbb{C}$ and its complex conjugate $\hat{k}_Q(x, x')^* = \langle \phi(x)|\phi(x')\rangle^* = \langle \phi(x')|\phi(x)\rangle$. Since the product of two kernels is known to be a valid kernel, it suffices to show that $\hat{k}_Q(x, x')$ is a valid complex-valued kernel and satisfies positive definiteness. For any $x^{(i)} \in \mathcal{X}, i = 1, \cdots, n$, and any coefficients $c_i \in \mathbb{C}$, the following holds

$$\sum_{i,j} c_i c_j^* \left(\hat{k}_Q(x^{(i)}, x^{(j)})\right) = \sum_{i,j} c_i c_j^* \langle \phi(x^{(i)})|\phi(x^{(j)})\rangle$$

$$= \left(\sum_i c_i \langle \phi(x^{(i)})|\right) \left(\sum_j c_j^* |\phi(x^{(j)})\rangle\right)$$

$$= \left\|\sum_i c_i^* |\phi(x^{(i)})\rangle\right\|^2 \geq 0. \tag{3.24}$$

This inequality confirms that $\hat{k}_Q(x, x')$ satisfies Mercer's condition to be a valid kernel as illustrated in Eq. (3.17). Therefore, the quantum kernel $k_Q(x, x')$ is also a valid kernel.

The inner product between quantum states can be efficiently estimated on quantum computers using techniques such as Loschmidt echo test [2] and SWAP test [3]. Both methods correspond to distinct quantum circuit architectures, as illustrated in Fig. 3.2.

One key merit of quantum kernels is that their derivation does not require the explicit representation of the quantum feature maps. Instead, it relies only on the

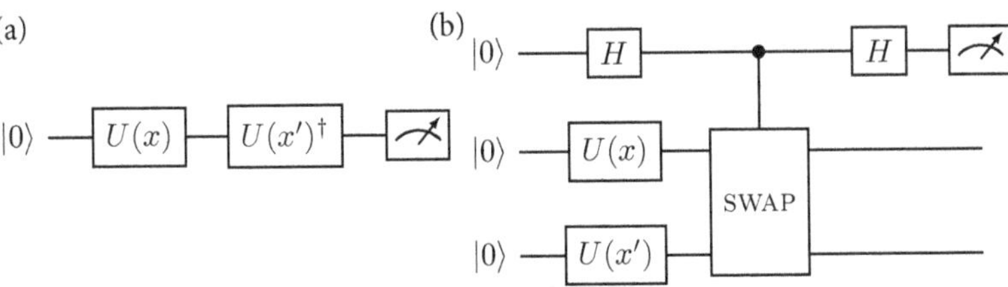

Fig. 3.2 Two methods for computing the inner product of the kernel. (**a**) Loschmidt echo test. (**b**) Swap test

construction of the associated quantum circuits. This aligns with the essence of kernel methods: while feature mappings can be computationally complex, the kernel function itself must remain efficient to evaluate.

Below, the core steps for constructing a quantum kernel are outlined.

General Construction Rules of Quantum Kernels

There are three steps to construct a quantum kernel:

1. Quantum feature map construction. Design a data-dependent quantum circuit $U(x)$ to encode classical input data x into the amplitudes or parameters of a quantum state $|\phi(x)\rangle = U(x)|\psi\rangle$ where the initial state $|\psi\rangle$ is typically $|0\rangle^{\otimes N}$.
2. Kernel evaluation. The quantum kernel is typically defined as the inner product of quantum states corresponding to two data points. Mathematically, this can be expressed as $k_Q(x, x') = |\langle \phi(x)|\phi(x')\rangle|^2$.
3. Post-processing. After executing the quantum circuit for different input pairs, measure the output and calculate the kernel matrix. This matrix will then be used in machine learning models, such as SVM.

Below is a simple example of a quantum kernel with an angle encoding feature map introduced in Sect. 2.3.1.

Example 3.4 (Single-Qubit Kernel) Consider an embedding that encodes a scalar input $x \in \mathbb{R}$ into the quantum state of a single qubit. The embedding is implemented by the Pauli-X rotation gate $RX(x) = e^{-ix\sigma_x/2}$, where σ_x is the Pauli-X operator. The quantum feature map is then given by $\phi : x \rightarrow \rho(x) = |\phi(x)\rangle \langle \psi(x)|$ with

$$
\begin{aligned}
|\phi(x)\rangle &= e^{-ix\sigma_x/2}|0\rangle \\
&= (\cos(x)\mathbb{I} - i \sin(x)\sigma_x)|0\rangle \\
&= \cos(x)|0\rangle - i \sin(x)|1\rangle,
\end{aligned} \tag{3.25}
$$

and hence the quantum kernel yields

$$
k(x, x') = \left| \cos\left(\frac{x}{2}\right) \cos\left(\frac{x'}{2}\right) + \sin\left(\frac{x}{2}\right) \sin\left(\frac{x'}{2}\right) \right|^2 = \cos\left(\frac{x - x'}{2}\right)^2,
$$

which is a translation invariant squared cosine kernel.

3.2.2 *Comparative Analysis: Quantum vs. Classical Kernel Methods*

Comparing the fundamental components of classical and quantum kernel machines provides an intuitive way to understand their connections and differences. As illustrated in Fig. 3.3, both classical and quantum kernels embed data points from the data space $\mathcal{X}$ into a high-dimensional space and then compute the kernel as the inner product of feature maps. The quantum kernel achieves this using quantum circuits, as indicated by the blue color. Table 3.1 summarizes how these components are implemented in classical and quantum kernel machines.

The main distinctions between classical and quantum kernel machines lie in the computation processes for feature mapping and kernel matrix construction, as outlined below:

- *Classical versus quantum feature maps.* Quantum feature maps encode data into quantum states, resulting in exponentially large complex-valued vectors $|\phi(\boldsymbol{x})\rangle \in \mathbb{C}^{2^N}$. Nevertheless, classical feature maps operate in finite-dimensional real-valued spaces $\phi(\boldsymbol{x}) \in \mathbb{R}^D$. While quantum feature maps can theoretically

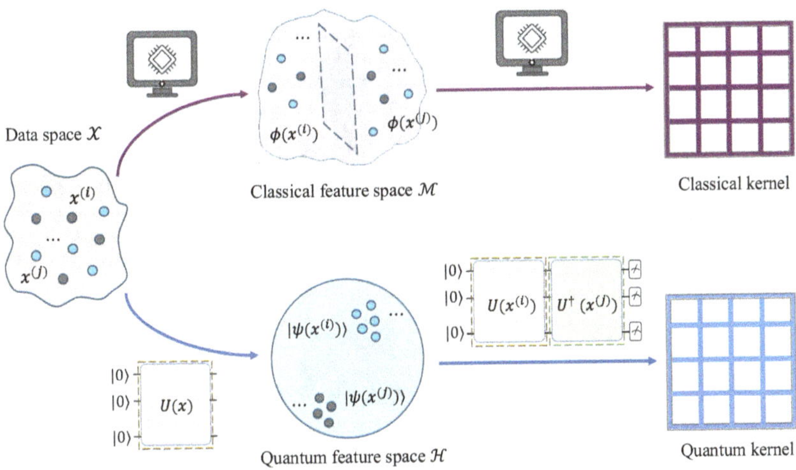

Fig. 3.3 The paradigm of classical and quantum kernels

Table 3.1 Comparison between classical and quantum kernel machines

	Classical kernel	Quantum kernel
Input	Classical data $\{\boldsymbol{x}^{(i)}\}_{i=1}^{n} \in \mathbb{R}^d$	Classical data $\{\boldsymbol{x}^{(i)}\}_{i=1}^{n} \in \mathbb{R}^d$
Feature	Real vector $\phi(\boldsymbol{x}) \in \mathbb{R}^D$	Complex vector $|\phi(\boldsymbol{x})\rangle \in \mathbb{C}^{2^N}$
Kernel	n-dimensional real matrix K_C	n-dimensional real matrix K_Q
Computation	Digital logical circuits $\phi(\boldsymbol{x})$	Quantum circuits $U(\boldsymbol{x})|\psi\rangle$

be simulated on classical computers by separating real and imaginary parts, this simulation becomes computationally infeasible as the number of qubits grows. Even feature maps generated by shallow quantum circuits are difficult to simulate efficiently on classical hardware, highlighting the inherent computational complexity of quantum feature maps.

- *Classical versus quantum kernels.* The kernel function is determined by the feature mapping, but its computational properties differ significantly between classical and quantum methods. A key merit of kernel methods is that they allow the use of complex feature maps while maintaining efficient kernel evaluations. Quantum kernels leverage quantum circuits to compute the inner product of quantum states, enabling the recognition of intricate patterns that classical kernels fail to capture. If a quantum kernel is computationally hard to evaluate classically, it offers a significant advantage by exploiting quantum computing's ability to process complex data representations efficiently.

> **Remark**
> The efficiency discussed here refers to the computational time within the respective classical or quantum frameworks as summarized below:
>
> - *Classical Efficiency*: Determined by the depth of digital logical circuits used for feature mapping and kernel computation.
> - *Quantum Efficiency*: Determined by the depth of quantum circuits required to achieve the same tasks.
>
> A computation process is considered efficient if it can be completed in polynomial time relative to the problem size in its corresponding framework (classical or quantum).

3.2.3 Concrete Examples of Quantum Kernels

To better understand the concept of a quantum kernel, common information encoding strategies used in quantum machine learning and their associated kernels are examined. It is important to note that some kernels cannot be efficiently computed on classical computers [4]. While such results are significant, the question of which quantum kernels are practically useful for real-world problems remains an open challenge.

In the following examples, the various encoding strategies introduced in Sect. 2.3.1 are first reviewed, and then, the corresponding quantum kernels are presented.

Example 3.5 (Quantum kernel with basis encoding) Given a classical binary vector $x = (x_1, \cdots, x_d) \in \{0, 1\}^d$, the quantum feature mapping related to basis encoding refers to

$$|\phi(x)\rangle = |x_1, \cdots, x_d\rangle, \tag{3.26}$$

and the induced quantum kernel yields

$$k(x, x') = |\langle \phi(x)|\phi(x')\rangle|^2 = \delta_{xx'}, \tag{3.27}$$

where $\delta_{xx'} = 1$ if $x = x'$ and otherwise 0.

The basis encoding requires $N = d$ qubits. This kernel function provides a very strict similarity measure on the input space and is arguably not the best choice of data encoding for quantum machine learning tasks.

Example 3.6 (Quantum kernel with amplitude encoding) Given a vector $x = (x_1, \cdots, x_d) \in \mathbb{R}^d$, the quantum feature mapping related to amplitude encoding refers to

$$|\phi(x)\rangle = \sum_{i=1}^{d} \frac{x_i}{\|x\|_2} |i\rangle, \tag{3.28}$$

where $\|x\|_2$ is the Euclidean norm. The related quantum kernel is given by

$$k(x, x') = |\langle \phi(x)|\phi(x')\rangle|^2 = \frac{|\langle x, x'\rangle|^2}{\|x\|_2^2 \cdot \|x'\|_2^2}. \tag{3.29}$$

The amplitude encoding requires $N = \lceil \log_2(d) \rceil$ qubits. This encoding strategy leads to an identity feature map, which can be implemented by a nontrivial quantum circuit (for obvious reasons also known as "arbitrary state preparation"), which takes time $O(d)$ in the worst case. Besides, this quantum kernel does not add much power to a linear model in the original feature space and is primarily of interest for theoretical investigations aimed at eliminating the effect of the feature map.

Example 3.7 (Quantum kernel with angle encoding) Given a vector $x = (x_1, \cdots, x_d) \in \mathbb{R}^d$, the quantum feature mapping related to angle encoding refers to

$$|\phi(x)\rangle = W_{d+1} e^{-ix_d G_d} W_d \cdots W_2 e^{-ix_1 G_1} W_1 |0\rangle^{\otimes d}, \qquad (3.30)$$

where $W_1, \cdots, W_{d+1}$ are arbitrary unitary evolutions and G_i is $d_i \leq d$-dimensional Hermitian operator called the generating Hamiltonian.

For a special case in which $W_i = \mathbb{I}$ and G_i refers to the Pauli-X operators σ_x acting on the i-th qubit, the quantum feature mapping refers to

$$|\phi(x)\rangle = \bigotimes_{i=1}^{d} \exp\left(-i\frac{x_i}{2}\sigma_x\right) |0\rangle^{\otimes d}, \qquad (3.31)$$

and the related quantum kernel is given by

$$k(x, x') = \prod_{i=1}^{d} \left| \sin(x_i)\sin(x_i') + \cos(x^{(i)})\cos(x_i') \right|^2$$

$$= \prod_{i=1}^{d} \left| \cos(x_i - x_i') \right|^2. \qquad (3.32)$$

The angle encoding requires d qubits, mapping the classical data to a 2^d-dimensional Hilbert space. A key advantage of angle encoding is its introduction of nonlinearity, a property essential for effective kernel-based machine learning. Specifically, nonlinearity enables the transformation of low-dimensional and non-linearly separable data into higher-dimensional, linearly separable representations. Additionally, angle encoding is well suited for implementation on modern devices with limited qubits and circuit depth, making it practical for exploring the utility of near-term quantum computers.

The quantum kernels related to different data encoding strategies have a resemblance to kernels from the classical machine learning literature. This means that sometimes up to an absolute square value, they can be identified with standard kernels such as the polynomial or Gaussian kernel. For the special case of angle encoding, the resemblance to classical kernels is because the employed quantum circuit does not employ any entangled quantum gates such that it can be simulated classically. We now discuss the general form of the quantum kernels induced by quantum feature maps from angle encoding in Eq. (3.30). The simplified case is considered where each input $x^{(i)}$ is encoded only once and all encoding Hamiltonians are the same, i.e., $G_1 = \cdots = G_d = G$.

Theorem 3.1 (Fourier Representation of the Quantum Kernel) *Let $X = \mathbb{R}^d$ and $U(x)$ be a quantum circuit that encodes the data inputs $x = (x_1, \cdots, x_d) \in X$ into a d-qubit quantum state $|\phi(x)\rangle$ via gates of the form $e^{-ix_i G}$ for $i = 1, \cdots, d$. Without loss of generality, G is assumed to be a $m \leq 2^d$-dimensional diagonal operator with spectrum $\lambda_1, \cdots, \lambda_m$. Between such data-encoding gates, and before and after the entire encoding circuit, arbitrary unitary evolutions $W_1, \cdots, W_{d+1}$ can be applied, so that*

$$U(x) = W_{d+1} e^{-ix_d G_d} W_d \cdots W_2 e^{-ix_1 G_1} W_1. \tag{3.33}$$

The quantum kernel $k_Q(x, x')$ can be written as

$$k_Q(x, x') = \sum_{s,t \in \Omega} e^{isx} e^{itx'} c_{st}, \tag{3.34}$$

where $\Omega \subset \mathbb{R}^d$ and $c_{st} \in \mathbb{C}$. For every $s, t \in \Omega$, we have $-s, -t \in \Omega$ and $c_{st} = c^*_{-s-t}$, which guarantees that the quantum kernel is real valued.

Proof Sketch of Theorem 3.1 The assumption that the generator G is diagonal could be made without loss of generality because Hermitian operators can be diagonalized as $G = V \Sigma V^\dagger$ with

$$e^{-ix_i \Sigma} = \begin{pmatrix} e^{-ix_i \lambda_1} & 0 & \cdots & 0 \\ 0 & e^{-ix_i \lambda_2} & \cdots & 0 \\ \cdots & \cdots & & \\ 0 & & \cdots & 0 \ e^{-ix_i \lambda_m} \end{pmatrix}, \tag{3.35}$$

where $V^\dagger$ refers to the conjugate transpose of the matrix V, $\lambda_1, \cdots, \lambda_m$ are the eigenvalues of G. Formally, $V, V^\dagger$ can be absorbed into the arbitrary circuits W_{i+1} and W_i before and after the encoding gate. In this regard, the quantum kernel can be written down as the inner product between the feature state of the specific forms in Eq. (3.30), i.e.,

$$k(x, x')$$

$$= \left| \langle \phi(x') | \phi(x) \rangle \right|$$

$$= \left| \langle 0 | W_1^\dagger (e^{-ix_1' \Sigma})^\dagger \cdots (e^{-ix_d' \Sigma})^\dagger W_{d+1}^\dagger W_{d+1} e^{-ix_d \Sigma} \cdots e^{-ix_1 \Sigma} W_1 |0\rangle \right|^2$$

$$= \left| \langle 0 | W_1^\dagger (e^{-ix_1' \Sigma})^\dagger \cdots (e^{-ix_d' \Sigma})^\dagger e^{-ix_d \Sigma} \cdots e^{-ix_1 \Sigma} W_1 |0\rangle \right|^2$$

$$= \left| \sum_{j_1, \cdots, j_d = 1}^{m} \sum_{k_1, \cdots, k_d = 1}^{m} e^{-i(\lambda_{j_1} x_1 - \lambda_{k_1} x_1' + \cdots + \lambda_{j_d} x_d - \lambda_{k_d} x_d')} \right.$$

Table 3.2 Overview of typical data encoding strategies and their quantum kernels. The input domain is assumed to be the $x = (x_1, \cdots, x_d) \in X \subset \mathbb{R}^d$

Encoding	Qubits	Dimension	Quantum kernel $k(x, x')$
Basis encoding	d	2^d	$\delta_{x,x'}$
Amplitude encoding	$\lceil \log_2(d) \rceil$	d	$|x^\dagger x'|^2$
Angle encoding	d	2^d	$\prod_{k=1}^d |\cos(x_k - x'_k)|^2$
General angle encoding	d	2^d	$\sum_{s,t \in \Omega} e^{-isx} e^{itx'} c_{st}$

$$
\times \left(W_1^{(1k_1)} \cdots W_d^{(k_{d-1}k_d)} \right)^* W_d^{(j_d j_{d-1})} \cdots W_1^{(j_1 1)} \Bigg|^2
$$

$$
= \left| \sum_j \sum_k e^{-i(\Lambda_j x - \Lambda_k x')} (\omega_k)^* \omega_j \right|^2
$$

$$
= \sum_j \sum_k \sum_h \sum_l e^{-i(\Lambda_j - \Lambda_l)x} e^{i(\Lambda_k - \Lambda_h)x'} (\omega_k \omega_h)^* \omega_j \omega_l, \tag{3.36}
$$

Here, the scalars $W_i^{(ab)}$ refer to the element $\langle a| W_i |b \rangle$ of the unitary operator W_i. The bold multi-index j represents the set $(j_1, \cdots, j_d)$, and Λ_j is a vector containing the eigenvalues selected by the multi-index (and similarly for k, h, l).

All terms where $\Lambda_j - \Lambda_l = s$ and $\Lambda_k - \Lambda_h = t$ can be summarized. In other words, the differences of eigenvalues amount to the same vectors s, t. Then

$$
k(x, x') = \sum_{s,t \in \Omega} e^{-isx} e^{itx'} \sum_{j,l:\Lambda_j - \Lambda_l = s} \sum_{k,h:\Lambda_k - \Lambda_h = t} \omega_j \omega_l (\omega_k \omega_h)^*
$$

$$
= \sum_{s,t \in \Omega} e^{-isx} e^{itx'} c_{st}. \tag{3.37}
$$

The frequency set Ω contains all vectors $\{\Lambda_j - \Lambda_l\}$ with $\Lambda_j = (\lambda_{j_1}, \cdots, \lambda_{j_d})$ and $\lambda_{j_1}, \cdots, \lambda_{j_d} \in [1, \cdots, m]$. $\qquad\qquad\square$

The various strategies for constructing quantum feature mappings and quantum kernels are summarized in Table 3.2.

Remark

After obtaining the quantum kernel matrix K_Q for a given training dataset $\{(x^{(i)}, y^{(i)})\}_{i=1}^n$, it can be used to perform regression or classification tasks in a manner similar to the classical kernel methods. In particular, as discussed in Sect. 3.1.2, consider the linear regression model given by

(continued)

$$\mathcal{L}(\boldsymbol{w}) = \frac{1}{2} \sum_{i=1}^{n} \left(\boldsymbol{w}^{\top} \cdot \phi_Q(\boldsymbol{x}^{(i)}) - y^{(i)} \right)^2 + \frac{\lambda}{2} \boldsymbol{w}^{\top} \cdot \boldsymbol{w}, \qquad (3.38)$$

where $\phi_Q(\boldsymbol{x}^{(i)})$ denotes the quantum feature mapping related to the quantum kernel k_Q, $\boldsymbol{w}$ denotes the model parameters, and $\lambda \geq 0$ is the regularization factor. Using the quantum kernel matrix, the dual representation of the linear model given in Eq. (3.14) can be expressed to predict the output for a new input as

$$y(\boldsymbol{x}) = \boldsymbol{k}_Q(\boldsymbol{x})^{\top} \cdot (K_Q + \lambda \mathbb{I}_n)^{-1} \boldsymbol{y}, \qquad (3.39)$$

where $\boldsymbol{y} = (y^{(1)}, \cdots, y^{(n)})$ refers to the label vector and $\boldsymbol{k}_Q(\boldsymbol{x})$ is a vector with elements $k_Q^{(i)}(\boldsymbol{x}) = k_Q(\boldsymbol{x}^{(i)}, \boldsymbol{x})$.

3.3 Theoretical Foundations of Quantum Kernel Machines

This section delves into the theoretical foundations of quantum kernels. Specifically, it focuses on two crucial aspects: the *expressivity* and *generalization* properties of quantum kernel machines. As shown in Fig. 3.4, these two aspects are essential for understanding the potential advantages of quantum kernels over classical learning approaches and their inherent limitations. In particular, expressivity refers to the size of the hypothesis space $\mathcal{H}_Q$ represented by quantum kernels, where $\mathcal{H}$ refers to the whole hypothesis space. Generalization considers the learned hypothesis that can accurately predict unseen data, exhibiting a small distance from the target concept. For ease of understanding, this section emphasizes the fundamental

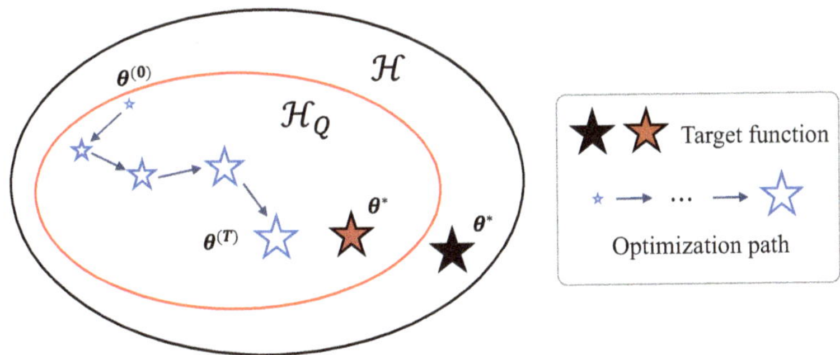

Fig. 3.4 The expressivity and generalization of quantum kernels

concepts necessary for evaluating the power and limitation of quantum kernels instead of exhaustively reviewing all theoretical results.

The outline of this chapter is as follows. In Sect. 3.3.1, the expressivity of quantum kernels is discussed, referring to the diversity of feature spaces that quantum kernels can represent. The insights gained will help identify tasks particularly well suited for quantum kernels. Then, in Sect. 3.3.2, the potential advantage of quantum kernels in terms of generalization error compared to all classical kernel machines is examined. This analysis highlights their ability to accurately predict labels or values for unseen data.

3.3.1 *Expressivity of Quantum Kernel Machines*

Quantum kernels, as discussed in Sect. 3.2, are constructed by explicitly defining quantum feature mappings. In this context, the expressivity of quantum kernel machines refers to the types of functions that quantum feature mappings can approximate and the kinds of correlations that quantum kernels can effectively model.

Following the conventions of [5], it is demonstrated that *any kernel function can be approximated using finitely deep quantum circuits* by showing that the associated feature mapping can also be approximated using quantum circuits. This conclusion rests on two key theoretical foundations: Mercer's feature space construction and the universality of quantum circuits. Together, these principles establish the theoretical feasibility of realizing any kernel function as a quantum kernel.

It is important to note that if exact mathematical equality were required, Mercer's construction would demand an infinite-dimensional Hilbert space, which in turn would require quantum computers with infinitely many qubits—an impractical scenario. However, in practical applications, approximating functions to a certain level of precision is more important than achieving exact evaluations. This perspective makes it feasible to implement the corresponding quantum feature mappings using a finite number of qubits. The following theorem confirms that any kernel function can be approximated as a quantum kernel to arbitrary precision with finite computational resources (proof details are deferred to the end of this subsection).

Theorem 3.2 (Approximate Universality of Finite-Dimensional Quantum Feature Maps) *Let $k : X \times X \to \mathbb{R}$ be a kernel function. Then, for any $\varepsilon \geq 0$, there exists $N \in \mathbb{N}$ and a quantum feature mapping ρ_N onto the Hilbert space of quantum states of N qubits such that*

$$|k(x, x') - 2^N \, Tr(\rho_N(x)\rho_N(x)') + 1| < \varepsilon \tag{3.40}$$

for almost all $x, x' \in X$.

Theorem 3.2, instead of discussing the ε-approximation of quantum kernels in the form $|k(x, x') - \mathrm{Tr}(\rho_N(x)\rho_N(x)')| < \varepsilon$, introduces additional multiplicative and additive factors. The corresponding mathematical expression is $|k(\boldsymbol{x}, \boldsymbol{x}') - 2^N \operatorname{Tr}(\rho_N(\boldsymbol{x})\rho_N(\boldsymbol{x}')) + 1| < \varepsilon$. These additional factors, explained below, do not impede the universality of the theorem.

Moreover, the statement that Eq. (3.40) holds for almost all $\boldsymbol{x}, \boldsymbol{x}' \in \mathcal{X}$ stems from measure theory. It signifies that the inequality is valid "except on sets of measure zero" or equivalently "with probability 1." In other words, while adversarial instances of $\boldsymbol{x}, \boldsymbol{x}' \in \mathcal{X}$ may exist for which the inequality does not hold, such instances are so sparse that the probability of encountering them when sampling from the relevant probability distribution is zero.

Last, Theorem 3.2 establishes that any kernel function can be approximated as a quantum kernel up to a multiplicative and an additive factor using a finite number of qubits.

Before presenting the proof of this theorem, Algorithm 1 is first introduced, which maps classical vectors to quantum states. These quantum states can then be used to evaluate Euclidean inner products as quantum kernels. Subsequently, we demonstrate Lemmas 3.1 and 3.2. These two lemmas separately formalize the correctness and runtime complexity of Algorithm 1, as well as establish the relationship between the Euclidean inner product of encoded real vectors and the Hilbert-Schmidt inner product of the corresponding quantum states.

Algorithm 1 Classical to quantum embedding (C2QE)

Input: a unit vector with 1-norm $\boldsymbol{r} \in \ell_1^d$
Output: Quantum state $\rho_r \propto \mathbb{I} + \sum_{i=1}^d r_i P_i$ ▷ See Lemma 3.1

1: Set $N = \lceil \log_4(d+1) \rceil$
2: Pad $\boldsymbol{r}$ with zeros until its length is $4^N - 1$
3: Draw $i \in \{1, \dots, 4^N - 1\}$ with probability $|r_i|$
4: Prepare $\rho_i = \frac{1}{2^N}(\mathbb{I} + \operatorname{sign}(r_i)P_i)$.
5: **return** ρ_i

The output of Algorithm 1, $\frac{1}{2^N}(\mathbb{I} \pm P)$, is a single (pure) eigenstate of a Pauli operator P with eigenvalue ± 1. However, since Line 3 involves sampling an index $i \in \{1, \cdots, 4^N - 1\}$, Algorithm 1 is inherently random, and the resulting quantum state is a classical mixture of pure states.

Lemma 3.1 (Correctness and Runtime of Algorithm 1) *Let $\boldsymbol{r} \in \ell_1^d \subset \mathbb{R}^d$ be a unit vector with respect to the 1-norm, i.e., $\|\boldsymbol{r}\|_1 = 1$. Take $N = \lceil \log_4(d + 1) \rceil$ and pad $\boldsymbol{r}$ with zeros until its length is $4^N - 1$. Let $(P_i)_{i=1}^{4^N-1}$ be the set of all Pauli matrices on N qubits, excluding the identity. Then, Algorithm 1 prepares the following state as a classical mixture:*

$$\rho(\cdot) : \ell_1^d \to Herm(2^N),$$

$$\boldsymbol{r} \mapsto \rho_r = \frac{\mathbb{I} + \sum_{i=1}^{4^N-1} r_i P_i}{2^N}. \tag{3.41}$$

The total runtime complexity t of Algorithm 1 fulfills $t \leq O(poly(d))$.

Proof of Lemma 3.1 The proof begins by expanding the state as follows:

$$\frac{\mathbb{I} + \sum_{i=1}^{4^N-1} r_i P_i}{2^N} = \frac{1}{2^N}\left(\sum_{i=1}^{4^N-1} |r_i|\mathbb{I} + \sum_{i=1}^{4^N-1} r_i P_i\right), \tag{3.42}$$

where the first equality follows that $\|r\|_1 = 1$ and $r \in \mathbb{R}^{4^N-1}$. Rewriting the above equation using $\mathrm{sign}(r_i)$ yields

$$\frac{\mathbb{I} + \sum_{i=1}^{4^N-1} r_i P_i}{2^N} = \frac{1}{2^N}\left(\sum_{i=1}^{4^N-1} |r_i|\mathbb{I} + \sum_{i=1}^{4^N-1} |r_i|\mathrm{sign}(r_i) P_i\right)$$

$$= \frac{1}{2^N}\sum_{i=1}^{4^N-1} |r_i|\,(\mathbb{I} + \mathrm{sign}(r_i) P_i) \succeq 0. \tag{3.43}$$

This uses the fact that $\sum_i |r_i| = \|r\|_1 = 1$ and $\mathbb{I} \pm P_i \geq 0$ for all Pauli operators P_i. Efficiently preparing $\mathbb{I} + P_i$ is achieved by rotating each qubit's $|0\rangle$ basis state to the corresponding Pauli basis and flipping the necessary qubits individually. Since this state is a convex combination of quantum states, it can be efficiently prepared by mixing, when the number of terms is polynomial. $\qquad\square$

Lemma 3.2 (Euclidean Inner Products) *Let $r, r' \in \mathbb{R}^d$ be unit vectors with respect to the 1-norm, i.e., $\|r\|_1 = \|r'\|_1 = 1$. For the states ρ_r, ρ'_r produced in Algorithm 1, the following identity holds*

$$\langle r, r'\rangle = 2^N\, Tr(\rho_r \rho'_r) - 1. \tag{3.44}$$

Proof of Lemma 3.2 The proof utilizes the following principles: (1) The trace is linear, and the trace of a tensor product equals the product of traces. (2) All Pauli words are traceless except for the identity, and each Pauli operator is its own inverse. Hence, the product of distinct Pauli operators is also traceless.

Expanding the trace of $\rho_r \rho_{r'}$, we have

$$\mathrm{Tr}(\rho_r \rho_{r'}) = \mathrm{Tr}\left(\frac{1}{4^N}\left(\mathbb{I} + \sum_{j=1}^{4^N-1} r_j P_j\right)\left(\mathbb{I} + \sum_{k=1}^{4^N-1} r'_k P_k\right)\right), \tag{3.45}$$

which could be simplified as

$$\mathrm{Tr}(\rho_r \rho_{r'}) = \frac{1}{4^N}\,\mathrm{Tr}\left(\mathbb{I} + \sum_{j=1}^{4^N-1} r_j r'_j P_j^2 + \sum_{k\neq j} r_j r'_k P_j P_k\right). \tag{3.46}$$

Using the properties of Pauli operators, the trace becomes

$$\text{Tr}(\rho_r \rho_{r'}) = \frac{1}{4^N} \left(\text{Tr}\,(\mathbb{I}) + \text{Tr} \left(\sum_{j=1}^{4^N-1} r_j r'_j \mathbb{I} \right) \right) = \frac{1 + \langle r, r' \rangle}{2^N}. \tag{3.47}$$

This completes the proof. $\qquad\square$

Lemma 3.2 clarifies the origin of the extra factors in Theorem 3.2. In particular, the 2^N multiplicative factor is unproblematic, as $N \leq O(\log(d))$ and the methods are designed to scale polynomially with d. Moreover, the quantum state ρ_r is generally mixed but can be efficiently prepared. The mapping is injective but not surjective.

With these results in place, the proof of Theorem 3.2 is now presented.

Proof of Theorem 3.2 The proof follows from a corollary of Mercer's theorem and the universality of quantum computing. First, the corollary of the Mercer's Theorem (i.e., Fact 3.1) is employed, which states that an arbitrary kernel k admits a uniformly convergent expansion of the form in Eq. (3.16). This ensures the existence of a finite-dimensional feature map $\Phi_m : \mathcal{X} \to \mathbb{R}^m$ such that

$$\left| k(x, x') - \langle \Phi_m(x), \Phi_m(x') \rangle \right| < \varepsilon. \tag{3.48}$$

Without loss of generality, it is assumed that $\|\Phi_m(x)\| = 1$ for all $x \in \mathcal{X}$. The quantum state ρ_{Φ_m} can then be prepared, which requires $\lceil \log_4(m+1) \rceil$ qubits. By preparing two such states—one for $\Phi_m(x)$ and one for $\Phi_m(x')$—their inner product can be computed as the Hilbert-Schmidt inner product of the quantum states, as shown in Lemma 3.2. This leads to

$$\langle \Phi_m(x), \Phi_m(x') \rangle = 2^N \,\text{Tr}\left(\rho_{\Phi_m(x)} \rho_{\Phi_m(x')} \right) - 1. \tag{3.49}$$

For reference, note that $\text{Tr}\left(\rho_{\Phi_m(x)} \rho_{\Phi_m(x')} \right)$ can be computed using the SWAP test to an additive precision determined by the number of measurement shots. This allows approximation of the result efficiently to any desired polynomial additive precision. Consequently, it follows that

$$\left| k(x, x') - 2^N \,\text{Tr}\left(\rho_{\Phi_m(x)} \rho_{\Phi_m(x')} \right) + 1 \right| < \varepsilon, \tag{3.50}$$

for almost all $x, x' \in \mathcal{X}$. This completes the proof. $\qquad\square$

Theorem 3.2 does not aim to demonstrate any quantum advantage but rather establishes the ultimate expressivity of quantum kernels. The theorem guarantees the existence of a quantum kernel using a finite number of qubits. However, it does not specify the scaling of required qubits with increasing computational complexity of the kernel function k or with decreasing approximation error $\varepsilon > 0$. The number of qubits N will depend on certain properties of the kernel k and the approximation error ε. For instance, if the required number of qubits scales exponentially with these parameters, Theorem 3.2 would have limited practical utility. lso to be considered is the time required to find such a quantum kernel approximation, independently of memory and runtime requirements for preparing feature vectors and computing their inner product.

> **Remark**
> Although Theorem 3.2 establishes that all kernel functions can be realized as quantum kernels, there may still exist kernel functions that cannot be *realized efficiently* as quantum kernels. This observation requires us to identify quantum kernels that can be computed efficiently on quantum computers, i.e., in polynomial time.

3.3.2   *Generalization of Quantum Kernel Machines*

Generalization, which quantifies the ability of learning models (both classical and quantum) to predict unseen data, is a critical metric for evaluating the quality of a learning model. Due to its importance, this section analyzes the potential advantages of quantum kernels in terms of generalization.

To provide a comprehensive understanding, this section first elucidates the generalization error bounds for general kernel machines, establishing a unified framework for a fair comparison between quantum kernels and classical kernels. Subsequently, a geometric metric is introduced to assess the potential quantum advantage of quantum kernels with respect to generalization error for a fixed amount of training data.

3.3.2.1   Generalization Error Bound for Kernel Machines

The optimal learning models based on specified kernel machines, which can be either classical or quantum, are reviewed, as discussed in Sect. 3.1.2. Suppose we have obtained n training examples $\{(x^{(i)}, y^{(i)})\}_{i=1}^{n}$ with $x^{(i)} \in \mathbb{R}^d$ and $y^{(i)} = f(x^{(i)}) \in \mathbb{R}$, where f is the target function. After training on this data, there exists a machine learning algorithm that outputs $h(x) = w^{\dagger}\phi(x)$, where $\phi(x) \in \mathbb{C}^D$ refers to the hidden feature map corresponding the classical/quantum kernel function $k(x^{(i)}, x^{(j)}) = K_{ij} = \phi(x^{(i)}) \cdot \phi(x^{(j)})$. More precisely, the mean square error is considered as the loss function for such a task:

$$\mathcal{L}(w, x) = \lambda w^{\dagger} w + \sum_{i=1}^{n} \left(w^{\dagger}\phi(x^{(i)}) - y^{(i)} \right)^2, \tag{3.51}$$

where $\lambda \geq 0$ is the regularization parameters for avoiding overfitting.

The optimal parameters for optimizing this loss function are given by

$$w^* = \arg \min_{w \in \Theta} \mathcal{L}(w, x). \tag{3.52}$$

As discussed in Sect. 3.1.2, the optimal solution w^* in Eq. (3.52) has the explicit form of

$$w^* = \Phi^{\dagger}(K + \lambda \mathbb{I}_n)^{-1} y = \sum_{i=1}^{n} \sum_{j=1}^{n} \phi(x^{(i)})((K + \lambda \mathbb{I}_n)^{-1})_{ij} y^{(j)}, \tag{3.53}$$

where $y = [y^{(1)}, ..., y^{(n)}]^{\top}$ refers to the vector of labels and $K \in \mathbb{R}^{n \times n}$ is the kernel matrix, and the second equality follows that $\Phi = [\phi(x^{(1)}), \cdots, \phi(x^{(n)})]^{\dagger}$. Moreover, the norm of the optimal parameters has a simple form when $\lambda \to 0$, i.e.,

$$\|w^*\|_2^2 = y^{\top} K^{-1} y. \tag{3.54}$$

The prediction error of these learning models is

$$\epsilon_{w^*}(x) = \left| f(x) - (w^*)^{\dagger} \phi(x) \right|, \tag{3.55}$$

which is uniquely determined by the kernel matrix K and the hyperparameter λ as shown in Eq. (3.53). In particular, the focus will be on the upper bound on the expected prediction error, as the sum of training error and generalization error.

Prediction, Training, and Generalization Error
In the context of learning theory, the upper bound of the expected prediction error defined in Eq. (3.55) (a.k.a, expected risk) is achieved by separately analyzing the upper bounds of the training error (a.k.a, empirical risk) and the generalization error, i.e.,

$$\mathbb{E}_{x \sim \mathcal{D}} \epsilon_{w^*}(x) = \underbrace{\frac{1}{n} \sum_{i=1}^{n} \epsilon_{w^*}(x^{(i)})}_{\text{Training error}} + \underbrace{\mathbb{E}_{x \sim \mathcal{D}} \epsilon_{w^*}(x) - \frac{1}{n} \sum_{i=1}^{n} \epsilon_{w^*}(x^{(i)})}_{\text{Generalization error}}.$$

$$\tag{3.56}$$

This decomposition stems from the fact that data distribution $\mathcal{D}$ is inaccessible in most scenarios.

A rough derivation of the upper bound of training error and generalization error will now be presented, outlining the necessary steps for clarity, while omitting specific details found in [6].

- *Training error.* Employing the convexity of function and Jensen's inequality, the training error yields

$$\frac{1}{n}\sum_{i=1}^{n}\epsilon_{w^*}(x^{(i)}) \leq \sqrt{\frac{1}{n}\sum_{i=1}^{n}\left((w^*)^{\dagger}\phi(x^{(i)}) - y^{(i)}\right)^2}. \tag{3.57}$$

Moreover, combining with the expression for the optimal w^* given in Eq. (3.53), we can obtain the upper bound of training error in terms of the kernel matrix K and hyperparameter λ, i.e.,

$$\frac{1}{n}\sum_{i=1}^{n}\epsilon_{w^*}(x^{(i)}) \leq \sqrt{\frac{\lambda^2 y^{\top}(K + \lambda\mathbb{I}_n)^{-2}y}{n}}. \tag{3.58}$$

Note that when $\lambda = 0$ and K are invertible, the training error is zero. However, the hyperparameter is usually set to $\lambda > 0$ in practice.

- *Generalization error.* The derivation of generalization error is more complicated than training error, involving a basic theorem in statistic and learning theory, presented below.

Fact 3.2 (Theorem 3.3, [1]) *Let $\mathcal{G}$ be a family of function mappings from a set $\mathcal{Z}$ to $[0, 1]$. Then, for any $\delta > 0$, with probability at least $1 - \delta$ over identical and independent draw of n samples from $\mathcal{Z} : z^{(1)}, \cdots , z^{(n)}$, we have for all $g \in \mathcal{G}$*

$$\mathbb{E}_z g(z) \leq \frac{1}{n}\sum_{i=1}^{n}g(z^{(i)}) + 2\mathbb{E}_\sigma\left[\sup_{g\in\mathcal{G}}\frac{1}{n}\sum_{i=1}^{n}\sigma_i g(z^{(i)})\right] + 3\sqrt{\frac{\log(2/\delta)}{2n}}, \tag{3.59}$$

where $\sigma_1, \cdots , \sigma_n$ are in independent and uniform random variables over $\{1, -1\}$.

For kernel functions defined in Eq. (3.55), the set $\mathcal{Z}$ refers to the space of input vector with $z^{(i)} = x^{(i)}$ drawn from an input distribution. Each function g equals to ϵ_w/α for some w, where ϵ_w is defined in Eq. (3.55) and α is a normalization factor such that the range of ϵ_w/α is $[0, 1]$. Without loss of generality, assume $\alpha = 1$. For any specific parameter w, consider the special case of $\mathcal{G}$ with setting $\mathcal{G}_w = \{\epsilon_v \mid \forall \|v\| \leq \|w\|\}$. Then, the upper bound of generalization error for the optimal parameter is

$$\mathbb{E}_x\epsilon_{w^*}(x) - \frac{1}{n}\sum_{i=1}^{n}\epsilon_{w^*}(x^{(i)})$$

$$\leq 2\mathbb{E}_\sigma\left[\sup_{\|v\|\leq\|w^*\|}\frac{1}{n}\sum_{i=1}^{n}\sigma_i\epsilon_v(x^{(i)})\right] + 3\sqrt{\frac{\log(2\|w^*\|/\delta)}{2n}}. \tag{3.60}$$

Moreover, applying Talagrand's contraction lemma [1] to the first term on the right-hand side gives

$$\mathbb{E}_{\sigma}\left[\sup_{\|\boldsymbol{v}\|\leq\|\boldsymbol{w}^*\|}\frac{1}{n}\sum_{i=1}^{n}\sigma_i\epsilon_{\boldsymbol{v}}(\boldsymbol{x}^{(i)})\right]\leq\mathbb{E}_{\sigma}\left[\sup_{\|\boldsymbol{v}\|\leq\|\boldsymbol{w}^*\|}\frac{1}{n}\sum_{i=1}^{n}\sigma_i(\boldsymbol{w}^*)^{\dagger}\phi(\boldsymbol{x}^{(i)})\right]$$

$$\leq\sqrt{\frac{\|\boldsymbol{w}^*\|^2}{n}}, \tag{3.61}$$

where the first inequality follows that $\epsilon_{\boldsymbol{v}}(\boldsymbol{x}^{(i)})$ is Lipschitz continuous with respect to $(\boldsymbol{w}^*)^{\dagger}\cdot\phi(\boldsymbol{x}^{(i)})$ with Lipschitz constant 1, the second inequality follows direct algebra operation. For detailed simplification, refer to Lemma 1 of [6].

In conjunction with Eq. (3.60), Eq. (3.61), and the expression of the optimal parameter $\boldsymbol{w}^*$ given in Eq. (3.53), the final upper bound of generalization error in terms of the kernel matrix is derived:

$$\mathbb{E}_{\boldsymbol{x}}\epsilon_{\boldsymbol{w}^*}(\boldsymbol{x})-\frac{1}{n}\sum_{i=1}^{n}\epsilon_{\boldsymbol{w}^*}(\boldsymbol{x}^{(i)})$$

$$\leq 5\cdot\frac{\boldsymbol{y}^{\top}(K+\lambda\mathbb{I}_n)^{-1}K(K+\lambda\mathbb{I}_n)^{-1}\boldsymbol{y}}{n}+3\sqrt{\frac{\log(2/\delta)}{2n}}. \tag{3.62}$$

For the case of $\lambda=0$, the first term in the generalization error bound simplifies to $5\cdot\boldsymbol{y}^{\top}K^{-1}\boldsymbol{y}/n$.

With the upper bound of training and generalization error, the prediction error of the learning model for a specific kernel matrix can be directly obtained. These three errors are summarized below.

Remark

The upper bound of the prediction error for kernel methods defined in Eq. (3.55) refers to

$$\mathbb{E}_{\boldsymbol{x}\sim\mathcal{D}}\epsilon_{\boldsymbol{w}^*}(\boldsymbol{x})\leq O\Bigg(\underbrace{\sqrt{\frac{\lambda^2\boldsymbol{y}^{\top}(K+\lambda\mathbb{I}_n)^{-2}\boldsymbol{y}}{n}}}_{\text{Training error}}+$$

$$\underbrace{\sqrt{\frac{\boldsymbol{y}^{\top}(K+\lambda\mathbb{I}_n)^{-1}K(K+\lambda\mathbb{I}_n)^{-1}\boldsymbol{y}}{n}}+\sqrt{\frac{\log(1/\delta)}{n}}}_{\text{Generalization error}}\Bigg), \tag{3.63}$$

(continued)

where K is a specific kernel related to the learning models and $\mathbf{y} = [y^{(1)}, \cdots, y^{(n)}]$ refers to the label vector of n training data. For the special case of $\lambda = 0$, the training error is zero, and the prediction error reduces to the generalization error with a simple form

$$\mathbb{E}_{\mathbf{x} \sim \mathcal{D}} \epsilon_{\mathbf{w}^*}(\mathbf{x}) \le O\left(\sqrt{\frac{\mathbf{y}^\top K^{-1} \mathbf{y}}{n}} + \sqrt{\frac{\log(1/\delta)}{n}} \right). \tag{3.64}$$

Note that the derived upper bound of the prediction error applies to both classical and quantum kernels. This is because there are no restrictions imposed on the kernel matrix K during the derivation.

3.3.2.2 Quantum Kernels with Prediction Advantages

Using the above theoretical results of generalization error for general kernel machines, we now elucidate how to access the potential quantum advantage of quantum kernels. For a clear understanding, we focus on the case of $\lambda = 0$ in which the prediction error bound has a simple form of $O\left(\sqrt{\mathbf{y}^\top K^{-1} \mathbf{y}/n} + \sqrt{\log(1/\delta)/n}\right)$ as shown in Eq. (3.64). In particular, this bound has a key dependence on two quantities, namely, (1) the size of training data n and (2) the kernel-dependent term $\mathbf{y}^\top K^{-1} \mathbf{y}$, which we denote as

$$s_K(\mathbf{y}) = \mathbf{y}^\top K^{-1} \mathbf{y}, \tag{3.65}$$

in the following discussion for simplification.

The dependence on n reflects the role of data to improve prediction performance. On the other hand, the quantity $s_K(\mathbf{y})$ is equal to the model complexity of the trained function $h(\mathbf{x}) = (\mathbf{w}^*)^\dagger \cdot \phi(\mathbf{x})$, where $s_K(\mathbf{y}) = \|\mathbf{w}^*\|^2 = (\mathbf{w}^*)^\dagger \cdot \mathbf{w}^*$ after training. A smaller value of $s_K(\mathbf{y})$ implies better generalization to new data $\mathbf{x}$ sampled from the distribution $\mathcal{D}$. Intuitively, $s_K(\mathbf{y})$ measures whether the closeness between $\mathbf{x}^{(i)}$ and $\mathbf{x}^{(j)}$ defined by the kernel function $k(\mathbf{x}^{(i)}, \mathbf{x}^{(j)})$ matches well with the closeness of the labels $y^{(i)}$ and $y^{(j)}$, recalling that a larger kernel value indicates two points are closer.

Based on the above discussion, it is interesting to analyze the potential advantage of quantum kernel machines. Given a set of training data $\{(\mathbf{x}^{(i)}, y^{(i)})\}_{i=1}^{n}$, let Q and C be the class of **Quantum** and **Classical** kernels, respectively, that can be efficiently evaluated on quantum and classical computers for any given $\mathbf{x}$. In order to formally evaluate the potential for quantum prediction advantage generally, one must take the quantum kernel $K_Q \in Q$ to satisfy the following two conditions:

- K_Q is hard to compute classically for any given $\boldsymbol{x}$.
- According to Eq. (3.65), the quantity $s_Q(\boldsymbol{y})$ related to the quantum kernel K_Q must be the minimal over all efficient classical models, namely, $s_Q(\boldsymbol{y}) \leq s_C(\boldsymbol{y})$ for any $K_C \in C$ with $s_C(\boldsymbol{y})$ being the K_C related quantity.

From the second condition, one can see that the potential advantage for the quantum kernel K_Q to predict better than a classical kernel K_C depends on the largest possible separation between $s_Q(\boldsymbol{y})$ and $s_C(\boldsymbol{y})$ for a dataset. Huang et al. [6] define a geometry metric, namely, **asymmetric geometric difference**, to characterize this separation for a fixed training dataset, which is given by

$$
g_{CQ} = g(K_C \| K_Q) = \sqrt{\left\| \sqrt{K_Q}(K_C^{-1})\sqrt{K_Q} \right\|_\infty},
\tag{3.66}
$$

where $\| \cdot \|_\infty$ is the spectral norm of the resulting matrix and we assume $\mathrm{Tr}(K_Q) = \mathrm{Tr}(K_C) = n$. The geometric difference $g(K_C \| K_Q)$ can be computed on a classical computer by performing a singular value decomposition of the $n \times n$ matrices K_C and K_Q in time at most order n^3.

Figure 3.5 presents a detailed flowchart for evaluating the potential quantum prediction advantage using the defined geometric difference g_{CQ} in a machine learning task. The input consists of n data samples, along with both quantum and classical methods, each associated with its respective kernel. The tests are conducted as a function of n to highlight the role of data size in determining the potential for a prediction advantage.

First, the geometric quantity g_{CQ} is evaluated, which quantifies the potential for a separation between quantum and classical predictions, without yet considering the actual function to be learned. Specifically, a large value of $g_{CQ} \propto \sqrt{n}$ suggests the possibility of a quantum prediction advantage. If the test is passed, an adversarial dataset that saturates this limit can be constructed. In particular, there exists a dataset with $s_C = g_{CQ}^2 s_Q$, where the quantum model exhibits superior prediction performance, as will be described in the subsequent context.

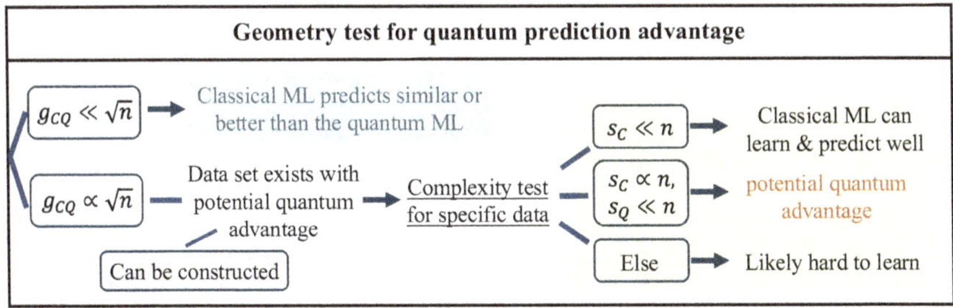

Fig. 3.5 A flowchart for understanding the potential for quantum prediction advantage (Adapted from [6])

Subsequently, to incorporate the provided data, a label-specific test can be performed using the model complexities s_C and s_Q. For quantum kernels and classical learning models, when $s_Q \ll n$ and $s_C \propto n$, a prediction advantage for quantum models is possible, as supported by the generalization bound in Eq. (3.64). In contrast, if g_{CQ} is small such as $g_{CQ} \ll \sqrt{n}$, the classical learning model will likely have a similar or better model complexity $s_C(y)$ compared to the quantum model. In this case, the classical model's prediction performance will be competitive or superior, and the classical model would likely be preferred.

3.3.2.3 Construction of Dataset with Maximal Quantum Advantage

The construction rule of a dataset that enables the maximal separation between the model complexity of quantum kernels and classical kernels is as follows. As indicated by the geometry test in Fig. 3.5, to separate between quantum and classical models related to kernel matrix K_Q and K_C, the ratio between s_C and s_Q should be as large as possible for a particular choice of targets $y^{(1)}, \cdots , y^{(n)}$. This could be achieved by solving the optimization problem:

$$\min_{y \in \mathbb{R}^n} \frac{s_C}{s_Q} = \min_{y \in \mathbb{R}^n} \frac{y^\top K_C^{-1} y}{y^\top K_Q^{-1} y}, \tag{3.67}$$

which has an exact solution given by a generalized eigenvalue problem. The solution is given by $y = \sqrt{K_Q} v$, where v is the eigenvector of $\sqrt{K_Q} K_C^{-1} \sqrt{K_Q}$ corresponding to the eigenvalue $g^2 = \|\sqrt{K_Q} K_C^{-1} \sqrt{K_Q}\|_\infty$. This guarantees that $s_C = g^2 s_Q$, and note that by definition of $g, s_C \leq g^2 s_Q$. Hence, this dataset fully utilized the geometric difference between the quantum and classical space. Finally, we can turn this dataset, which maps input x to a real value y_Q, into a classification task by replacing y_Q with $+1$ if $y_Q > \mathrm{median}(y^{(1)}, \cdots , y^{(n)})$ and -1 if $y_Q \leq \mathrm{median}(y^{(1)}, \cdots , y^{(n)})$. The constructed dataset will yield the largest separation between quantum and classical models from a learning theoretic sense, as the model complexity fully saturates the geometric difference. If there is no quantum advantage in this dataset, there will likely be none.

3.4 Code Demonstration

This section explores the practical implementation of a quantum kernel. Before diving into concrete code examples, an efficient strategy for estimating the quantum kernel in practice is discussed.

As explained in Sect. 3.2, one method for estimating the quantum kernel is the SWAP test, which is resource-intensive. An alternative is to encode the classical data vector x using a unitary operation $U(x)$ and apply the inverse embedding of x'

using $U(\boldsymbol{x}')^\dagger$ on the same qubits. The quantum kernel $k_Q(\boldsymbol{x}, \boldsymbol{x}')$ is then estimated by measuring the expectation of the projector $O = (|0\rangle \langle 0|)^{\otimes N}$ on the zero state.

The complete quantum circuit architecture for this process is illustrated in Fig. 3.3. Mathematically, the process is expressed as

$$\langle 0^{\otimes N}|U(\boldsymbol{x}')U(\boldsymbol{x})^\dagger O U(\boldsymbol{x}')^\dagger U(\boldsymbol{x})|0^{\otimes N}\rangle$$

$$= \langle 0^{\otimes N}|U(\boldsymbol{x}')U(\boldsymbol{x})^\dagger|0\rangle^{\otimes N} \langle 0|^{\otimes N} U(\boldsymbol{x}')^\dagger U(\boldsymbol{x})|0^{\otimes N}\rangle$$

$$= \left|\langle 0^{\otimes N}|U(\boldsymbol{x}')^\dagger U(\boldsymbol{x})|0^{\otimes N}\rangle\right|^2$$

$$= \left|\langle \phi(\boldsymbol{x})|\phi(\boldsymbol{x}')\rangle\right|^2$$

$$= k_Q(\boldsymbol{x}, \boldsymbol{x}'). \tag{3.68}$$

This approach allows the quantum kernel estimation to use the same number of qubits required for the quantum feature mapping of the classical vector $\boldsymbol{x}$.

Next, an example demonstrating the workflow of applying quantum kernels for classification tasks on the MNIST dataset is provided, with step-by-step code implementation.

3.4.1 Classification on MNIST Dataset

We train a support vector machine (SVM) classifier associated with a quantum kernel on the MNIST dataset, a widely used benchmark in image classification. To assess the performance of the quantum kernel-based classifier, we adopt the classification accuracy, a standard metric in classification tasks. That is, the classification accuracy is defined as the proportion of correctly classified samples out of the total number of samples.

The pipeline involves the following steps:

Step 1 Load and preprocess the dataset.
Step 2 Define the quantum feature mapping.
Step 3 Construct the quantum kernel.
Step 4 Train and evaluate the SVM classifier.

We begin by importing the required libraries.

```python
import pennylane as qml
from sklearn.datasets import fetch_openml
from sklearn.decomposition import PCA
from sklearn.model_selection import train_test_split
from sklearn.preprocessing import StandardScaler
from sklearn.svm import SVC
from sklearn.metrics import accuracy_score
import numpy as np
```

Step 1: Dataset Preparation The focus is on the digits 3 and 6 in the MNIST dataset, forming a binary classification problem. Principal component analysis (PCA) [7] is applied to reduce the feature dimension of the images, minimizing the number of required qubits for encoding. The compressed features are normalized to align with the periodicity of the quantum feature mapping.

```python
def load_mnist(n_qubit):
    # Load MNIST dataset from OpenML
    mnist = fetch_openml('mnist_784', version=1)
    X, y = mnist.data, mnist.target

    # Filter out the digits 3 and 6
    mask = (y == '3') | (y == '6')
    X_filtered = X[mask]
    y_filtered = y[mask]

    # Convert labels to binary (0 for digit 3 and 1 for digit
        6)
    y_filtered = np.where(y_filtered == '3', 0, 1)

    # Apply PCA to reduce feature dimension
    pca = PCA(n_components=n_qubit)
    X_reduced = pca.fit_transform(X_filtered)

    # Normalize the input features
    scaler = StandardScaler().fit(X_reduced)
    X_scaled = scaler.transform(X_reduced)

    # Split into training and testing sets
    X_train, X_test, y_train, y_test = train_test_split(
        X_scaled, y_filtered, test_size=0.2, random_state=42)
    return X_train, X_test, y_train, y_test

n_qubit = 8
X_train, X_test, y_train, y_test = load_mnist(n_qubit)
```

To better understand the structure of the dataset, the training data are visualized using t-distributed stochastic neighbor embedding (t-SNE) [8]. The following code generates the visualization:

```python
def visualize_dataset(X, labels):
    import matplotlib.pyplot as plt
    from sklearn.manifold import TSNE

    tsne = TSNE(n_components=2, random_state=42, perplexity
        =30)
    label2name = {
        0: '3',
        1: '6'
    }
    mnist_tsne = tsne.fit_transform(X)
    for label in np.unique(labels):
        indices = labels == label
```

```
13    plt.scatter(mnist_tsne[indices, 0], mnist_tsne[indices
          , 1], cmap='coolwarm', s=20, label=f'Number_{
          label2name[label]}')
14
15    # Add labels and legend
16    plt.title("t-SNE_Visualization_of_Two_Classes_(3_and_6)")
17    plt.xlabel("t-SNE_Dimension_1")
18    plt.ylabel("t-SNE_Dimension_2")
19    plt.legend()
20
21    plt.tight_layout()
22    plt.show()
23
24 visualize_dataset(X_train, y_train)
```

The resulting t-SNE visualization is shown in Fig. 3.6.

Steps 2 and 3: Define Quantum Feature Mapping and Building Quantum Kernel We use angle embedding as the quantum feature mapping method. The quantum kernel is implemented as follows.

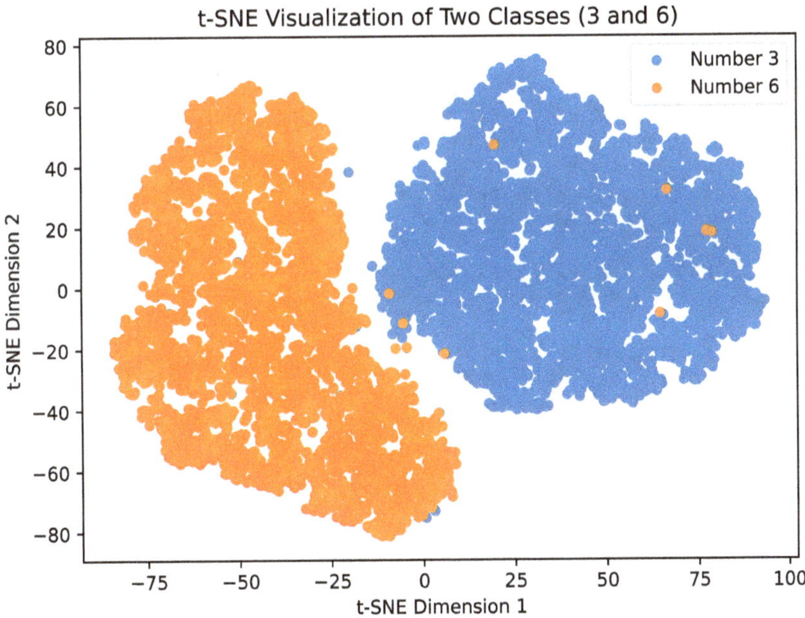

Fig. 3.6 T-SNE visualization of MNIST dataset of two classes "3" and "6"

```
dev = qml.device('default.qubit', wires=n_qubit)

@qml.qnode(dev)
def kernel(x1, x2, n_qubit):
    qml.AngleEmbedding(x1, wires=range(n_qubit))
    qml.adjoint(qml.AngleEmbedding)(x2, wires=range(n_qubit))
    return qml.expval(qml.Projector([0]*n_qubit, wires=range(
        n_qubit)))
```

Using the quantum kernel, the kernel matrix is constructed by computing the kernel values for all pairs of samples:

```
def kernel_mat(A, B):
    mat = []
    for a in A:
    row = []
    for b in B:
    row.append(kernel(a, b, n_qubit))
    mat.append(row)
    return np.array(mat)
```

Next, the quantum kernel matrix is visualized to gain insight into its structure.

```
def visualize_kernel(X, y, n_sample):
    X_vis = []
    for label in np.unique(y):
        index = y == label
        X_vis.append(X[index][:n_sample])

    X_vis = np.concatenate(X_vis, axis=0)
    n_sample_per_class = len(X_vis) // 2

    sim_mat = kernel_mat(X_vis, X_vis)
    np.save('code/chapter_4_kernel/sim_mat.npy', sim_mat)

    import matplotlib.pyplot as plt
    plt.imshow(sim_mat, cmap='viridis', interpolation='nearest
        ')

    # Add color bar to show the scale
    plt.colorbar(label='Similarity')

    plt.axhline(n_sample_per_class - 0.5, color='red',
        linewidth=1.5)  # Horizontal line
    plt.axvline(n_sample_per_class - 0.5, color='red',
        linewidth=1.5)  # Vertical line

    xticks = yticks = np.arange(0, len(X_vis))
    xtick_labels = [f"3-{i+1}" if i < n_sample_per_class else
        f"6-{i+1-n_sample_per_class}" for i in range(len(X_vis
        ))]
    ytick_labels = xtick_labels
```

```
26    plt.xticks(xticks, labels=xtick_labels, rotation=90,
          fontsize=8)
27    plt.yticks(yticks, labels=ytick_labels, fontsize=8)
28
29    # Title and axis labels
30    plt.title("Quantum_Kernel_Matrix")
31    plt.xlabel("Sample_Index")
32    plt.ylabel("Sample_Index")
33
34
35    plt.tight_layout()
36    plt.show()
37
38 visualize_kernel(X_train, y_train, 10)
```

The resulting kernel matrix is shown in Fig. 3.7.

From the visualization, we observe a clear block structure:

- Most of the elements in the top-left and bottom-right blocks, where samples belong to the same class, show higher similarity values.
- Most of the elements in the top-right and bottom-left blocks, where samples belong to different classes, exhibit lower similarity values.

This indicates that it may be possible to distinguish the two classes by setting a similarity threshold.

Step 4: Training SVM We construct a SVM classifier with the quantum kernel matrix and train it with the training data prepared in Step 1:

```
1 svm = SVC(kernel=kernel_mat)
2 svm.fit(X_train, y_train)
3 pred = svm.predict(X_test)
4 print("Accuracy:", accuracy_score(y_test, pred))
```

Fig. 3.7 Visualization of quantum kernel matrix on 20 samples, equally drawn from two classes "3" and "6" in MNIST dataset

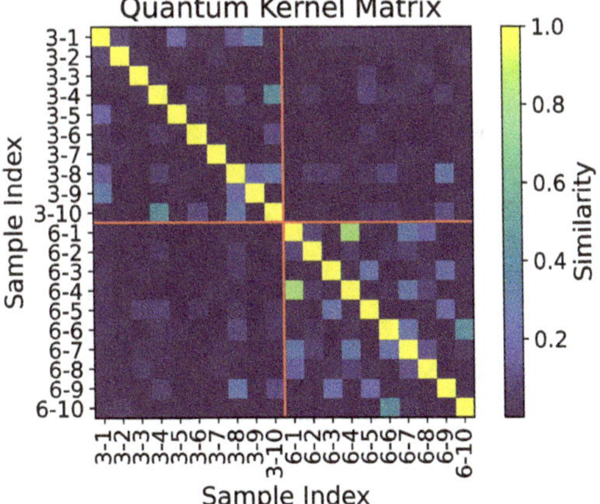

To further analyze how the performance of the SVM with a quantum kernel depends on the size of the training dataset, the number of training samples is varied from 10 to 100 in increments of 10. For each configuration, the corresponding classification accuracy on the test data is recorded.

```python
svm = SVC(kernel='precomputed')
n_sample_max = 100
X_train_sample = []
y_train_sample = []
for label in np.unique(y_train):
    index = y_train == label
    X_train_sample.append(X_train[index][:n_sample_max])
    y_train_sample.append(y_train[index][:n_sample_max])
X_train_sample = np.concatenate(X_train_sample, axis=0)
y_train_sample = np.concatenate(y_train_sample, axis=0)
kernel_mat_train = kernel_mat(X_train_sample, X_train_sample)
kernel_mat_test = kernel_mat(X_test, X_train_sample)

accuracy = []
n_samples = []
for n_sample in range(10, n_sample_max+10, 10):
    class1_indices = np.arange(n_sample)
    class2_indices = np.arange(n_sample_max, n_sample_max+
        n_sample)
    selected_indices = np.concatenate([class1_indices,
        class2_indices])

    svm.fit(kernel_mat_train[np.ix_(selected_indices,
        selected_indices)], np.concatenate([y_train_sample[:
        n_sample], y_train_sample[n_sample_max:n_sample_max+
        n_sample]]))
    pred = svm.predict(np.concatenate([kernel_mat_test[:, :
        n_sample], kernel_mat_test[:, n_sample_max:
        n_sample_max+n_sample]], axis=1))
    accuracy.append(accuracy_score(y_test, pred))
    n_samples.append(n_sample)

plt.plot(n_sample, accuracy, marker='o')
plt.title('Classification Accuracy vs. #Training Samples')
plt.xlabel('#Training Samples')
plt.xticks(n_sample, n_sample)
plt.ylabel('Accuracy')
plt.grid()
plt.tight_layout()
plt.show()
```

As illustrated in Fig. 3.8, the quantum kernel-based SVM achieves over 93% accuracy with just 20 training samples and continues to improve as more training data is provided, ultimately exceeding 99% accuracy with 200 training samples. This performance highlights the potential of quantum kernels in classification tasks.

To evaluate the effectiveness of quantum kernels in comparison to classical counterparts, an SVM classifier using three different classical kernels introduced

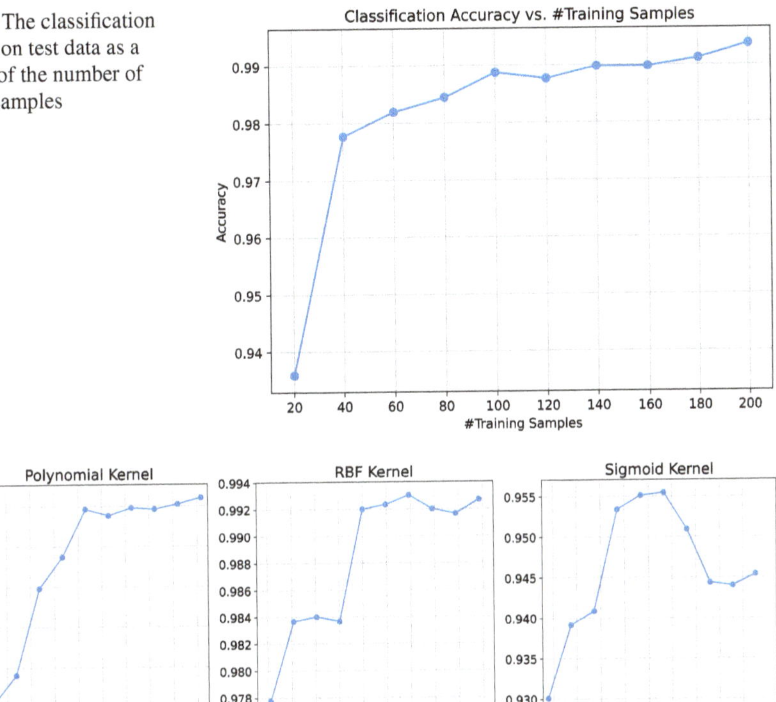

Fig. 3.8 The classification accuracy on test data as a function of the number of training samples

Fig. 3.9 The classification accuracy of SVM with classical kernels on test data as a function of the number of training samples

in this chapter is implemented, which are the polynomial kernel, the RBF kernel, and the sigmoid kernel. For a fair comparison, all experimental settings are kept identical, except for the choice of the kernel function. The hyperparameters for each classical kernel are set to their default values in the scikit-learn library. The same training and evaluation procedures used for the quantum kernel-based SVM are applied, and the results are summarized in Fig. 3.9. A comparison of these results reveals that the quantum kernel achieves performance comparable to that of the classical RBF kernel on this dataset. This suggests that quantum kernels can serve as a competitive alternative to well-established classical kernel methods for certain classical classification tasks.

3.5 Bibliographic Remarks

The foundational concept of using quantum computers to evaluate kernel functions, namely, the concept of quantum kernels, was first explored by Schuld et al. [9]. They highlighted the fundamental differences between quantum kernels and quantum support vector machines. Building on this, [10] and [11] established a connection between quantum kernels and parameterized quantum circuits (PQCs), demonstrating their practical implementation. These works emphasized the parallels between quantum feature maps and the classical kernel trick. Since then, a large number of studies delved into figuring out the potential of quantum kernels for solving practical real-world problems.

The recent advancements in quantum kernel machines can be roughly categorized into three key areas: *kernel design, theoretical findings*, and *applications*. Specifically, the advances in kernel design focus on addressing challenges such as vanishing similarity and kernel concentration by exploring innovative frameworks. Theoretical work studies how well quantum kernels might work on new data, how robust they are to noise, and whether they can really outperform classical methods. Practical studies focus on applying quantum kernels to real-world problems. The following sections discuss each of these areas.

3.5.1   *Quantum Kernel Design*

A crucial research line in this field focuses on constructing *trainable quantum kernels* to maximize performance for specific datasets and problem domains. In particular, traditional quantum kernels, with fixed data embedding schemes, are limited to specific feature representation spaces and often fail to capture the complex and diverse patterns inherent in real-world data. To tackle this, [12] designed feature maps tailored to each task. Later, [13] showed that quantum feature maps can be optimized—much like adjusting parameters in a neural network—using data re-uploading methods [14, 15]. Other researchers proposed combining different kernels or searching for the best quantum circuit architecture [16, 17]. Covariant quantum kernels, suggested by Glick et al. [18], are another approach for problems with group structure.

Another major challenge is the problem of "vanishing similarity," in quantum kernels [19], also called exponential kernel concentration [19]. Quantum kernels are based on the overlap between quantum feature maps. But in high-dimensional spaces, these feature maps tend to be almost orthogonal, making the kernels nearly useless for distinguishing data points [6]. As a result, models built on these kernels may not generalize well to new data.

To fix vanishing similarity, [6] suggested storing quantum features as classical vectors and using a standard Gaussian kernel. This bypasses the problem of near-orthogonal quantum states. Another method, the antisymmetric logarithmic derivative quantum Fisher kernel, encodes the geometric structure of the input data to avoid the same issue [20]. Others have tried rescaling the input data or tuning hyperparameters to keep feature maps closer together [21, 22], which helps prevent vanishing similarity but can make the kernels less flexible.

3.5.2 Theoretical Studies of Quantum Kernels

Theory work tries to find out how well quantum kernels really work, especially in practical conditions. Researchers focus on two big questions: how flexible quantum kernels are (can they model complex data?) and how well they generalize (do they work on new data, not just the training set?).

3.5.2.1 Expressivity of Quantum Kernels

The expressivity of quantum kernels refers to their capacity to capture complex data relationships and represent intricate patterns in the feature space. Researchers often use the idea of the reproducing kernel Hilbert space (RKHS) to analyze this, which tells us what kind of functions a kernel can model.

Schuld [23] rigorously analyzed the RKHS of embedding-based quantum kernels and established the universality approximation theorem, demonstrating that quantum kernels can approximate a wide class of functions. Building on this, [24] extended the analysis by investigating parameterized quantum embedding kernels, introducing a data-reuploading structure and proving a corresponding universality approximation theorem. These results underscore the expressive power of quantum kernels in representing complex data structures.

Even if a quantum kernel is flexible, it must also be efficient to build. If making a universal quantum kernel takes as long as classical methods, the advantage disappears.

To narrow this gap, [5] examined the expressive power of efficient quantum kernels that can be implemented on quantum computers within polynomial time. Their work provides a detailed analysis of the types of kernels that are achievable with a polynomial number of qubits and within polynomial time. The relevant results offer insights into the feasibility and practical utility of quantum kernels in real-world scenarios.

However, alongside the exploration of expressive power, a significant challenge known as exponential kernel concentration has been identified. Four things make this worse—too much expressivity in embeddings, global measurements, entanglement, and noise [19]. Many studies now focus on building new types of quantum kernels that avoid this pitfall, as discussed in Sect. 3.5.1.

3.5.2.2 Generalization of Quantum Kernels

Generalization—how well a model does on new data—is just as important as flexibility. Studies like [6] have set bounds for how much quantum kernels can generalize. For some quantum data, quantum kernels can learn patterns that classical models cannot.

However, quantum kernels often face practical hurdles like noise or limited measurements. For large datasets or when noise is strong, generalization can get much worse [25]. New methods like indefinite kernel learning can help keep performance strong, even in tough conditions.

Quantum kernels can also help for some hard classical problems. For example, with datasets based on the discrete logarithm, quantum kernels predict quickly, while classical models need much more time [4]. This hints at real speedup for some problems.

Still, quantum kernels are not always better. Without built-in "inductive bias"—rules that help models guess well on new data—they can underperform compared to classical models [26]. The way data is encoded matters a lot.

3.5.2.3 Provable Advantages of Quantum Kernels

The potential for quantum kernels to demonstrate quantum advantage has been a central focus of research. For instance, [6] provided evidence of generalization advantages for quantum kernels on quantum data. Similarly, [4] presented a rigorous framework showing that quantum kernels can efficiently solve problems like the discrete logarithm problem, which is believed to be intractable for classical computers under standard cryptographic assumptions. Moreover, [27] demonstrated quantum advantage in distribution learning tasks, offering some of the earliest theoretical evidence of quantum advantage in machine learning.

However, many of these tasks are artificial, designed specifically to showcase quantum advantages. This raises the question of how these theoretical benefits can be translated to real-world applications. In this regard, the next significant challenge is to demonstrate that quantum models can consistently outperform classical models in solving practical, real-world problems.

3.5.3 Applications of Quantum Kernels

Motivated by the potential of quantum kernels to recognize complex data patterns, numerous studies have explored their practical applications across diverse fields, including classification, drug discovery, anomaly detection, and financial modeling.

For instance, [28] investigate the use of quantum kernels for image classification, specifically in identifying real-world manufacturing defects. Similarly, [29] apply quantum kernels to satellite image classification, a task of particular importance in the earth observation industry. In the field of quantum physics, [30] and [31] leverage quantum kernels to recognize phases of quantum matter, where quantum kernels outperform classical learning models in solving certain problems. In drug discovery, [32] explore the potential of quantum kernels to accelerate and improve the identification of promising compounds.

Quantum kernels have also been explored in anomaly detection. Liu and Rebentrost [33] demonstrate their superior performance over classical methods in detecting anomalies within quantum data. Furthermore, [34] employ quantum kernel methods for fraud classification tasks, showing improvements when benchmarked against classical methods. Miyabe et al. [35] expand their application to the financial domain by proposing a quantum multiple-kernel learning methodology. This approach broadens the scope of quantum kernels to include credit scoring and directional forecasting of asset price movements, highlighting their potential utility in financial services.

Despite the promise of quantum kernels shown in specialized scenarios, their empirical advantages over classical models remain limited to specific problem settings, such as the negative impact of noise and large computation complexity in handling large-scale datasets. The realization of quantum advantage in practical tasks remains an ongoing area of research, with current efforts directed toward identifying real-world problems where quantum kernels outperform classical alternatives.

References

1. Mohri, M. (2018). *Foundations of machine learning*. MIT Press.
2. Kusumoto, T., Mitarai, K., Fujii, K., Kitagawa, M., & Negoro, M. (2021). Experimental quantum kernel trick with nuclear spins in a solid. *npj Quantum Information, 7*(1), 94.
3. Blank, C., Park, D. K., Rhee, J.-K. K., & Petruccione, F. (2020). Quantum classifier with tailored quantum kernel. *npj Quantum Information, 6*(1), 41.
4. Liu, Y., Arunachalam, S., & Temme, K. (2021). A rigorous and robust quantum speed-up in supervised machine learning. *Nature Physics, 17*(9), 1013–1017.
5. Gil-Fuster, E., Eisert, J., & Dunjko, V. (2024). On the expressivity of embedding quantum kernels. *Machine Learning: Science and Technology, 5*(2), 025003.
6. Huang, H.-Y., Broughton, M., Mohseni, M., Babbush, R., Boixo, S., Neven, H., McClean, J. R. (2021). Power of data in quantum machine learning. *Nature Communications, 12*(1), 2631.
7. Abdi, H., & Williams, L. J. (2010). Principal component analysis. *Wiley Interdisciplinary Reviews: Computational Statistics, 2*(4), 433–459.
8. Van der Maaten, L., & Hinton, G. (2008). Visualizing data using t-sne. *Journal of Machine Learning Research, 9*(11), 2579–2605.
9. Schuld, M., Fingerhuth, M., & Petruccione, F. (2017). Implementing a distance-based classifier with a quantum interference circuit. *Europhysics Letters, 119*(6), 60002.
10. Havlicek, V., Corcoles, A. D., Temme, K., Harrow, A. W., Kandala, A., Chow, J. M., & Gambetta, J. M. (2019). Supervised learning with quantum-enhanced feature spaces. *Nature, 567*(7747), 209–212.
11. Schuld, M., & Killoran, N. (2019). Quantum machine learning in feature hilbert spaces. *Physical Review Letters, 122*(4), 040504.
12. Lloyd, S., Schuld, M., Ijaz, A., Izaac, J., & Killoran, N. (2020). Quantum embeddings for machine learning. arXiv preprint arXiv:2001.03622.
13. Hubregtsen, T., Wierichs, D., Gil-Fuster, E., Derks, P.-J. H. S., Faehrmann, P. K., & Meyer, J. J. (2022). Training quantum embedding kernels on near-term quantum computers. *Physical Review A, 106*(4), 042431.
14. Pérez-Salinas, A., Cervera-Lierta, A., Gil-Fuster, E., & Latorre, J. I. (2020). Data re-uploading for a universal quantum classifier. *Quantum, 4*, 226.

15. Schuld, M., Sweke, R., Meyer, J. J. (2021). Effect of data encoding on the expressive power of variational quantum-machine-learning models. *Physical Review A, 103*(3), 032430.
16. Vedaie, S. S., Noori, M., Oberoi, J. S., Sanders, B. C., & Zahedinejad, E. (2020). Quantum multiple kernel learning. arXiv preprint arXiv:2011.09694.
17. Lei, C., Du, Y., Mi, P., Yu, J., & Liu, T. (2024). Neural auto-designer for enhanced quantum kernels. arXiv preprint arXiv:2401.11098.
18. Glick, J. R., Gujarati, T. P., Corcoles, A. D., Kim, Y., Kandala, A., Gambetta, J. M., & Temme, K. (2024). Covariant quantum kernels for data with group structure. *Nature Physics, 20*(3), 479–48.
19. Thanasilp, S., Wang, S., Cerezo, M., & Holmes, Z. (2022). Exponential concentration and untrainability in quantum kernel methods. arXiv preprint arXiv:2208.11060.
20. Suzuki, Y., Kawaguchi, H., & Yamamoto, N. (2022). Quantum fisher kernel for mitigating the vanishing similarity issue. *Quantum Science and Technology, 9*(3), 035050.
21. Shaydulin, R., & Wild, S. M. (2022) Importance of kernel bandwidth in quantum machine learning. *Physical Review A, 106*(4), 042407.
22. Canatar, A., Peters, E., Pehlevan, C., Wild, S. M., & Shaydulin, R. (2022). Bandwidth enables generalization in quantum kernel models. arXiv preprint arXiv:2206.06686.
23. Schuld, M. (2021). Supervised quantum machine learning models are kernel methods. arXiv preprint arXiv:2101.11020.
24. Jerbi, S., Fiderer, L. J., Nautrup, H. P., Kübler, J. M., Briegel, H. J., & Dunjko, V. (2023). Quantum machine learning beyond kernel methods. *Nature Communications, 14*(1), 1–8.
25. Wang, X., Du, Y., Luo, Y., & Tao, D. (2021). Towards understanding the power of quantum kernels in the nisq era. *Quantum, 5*, 531.
26. Kübler, J., Buchholz, S., & Schölkopf, B. (2021). The inductive bias of quantum kernels. *Advances in Neural Information Processing Systems, 34*, 12661–12673.
27. Sweke, R., Seifert, J.-P., Hangleiter, D., & Eisert, J. (2021). On the quantum versus classical learnability of discrete distributions. *Quantum, 5*, 417.
28. Beaulieu, D., Miracle, D., Pham, A., & Scherr, W. (2022). Quantum kernel for image classification of real world manufacturing defects. arXiv preprint arXiv:2212.08693.
29. Rodriguez-Grasa, P., Farzan-Rodriguez, R., Novelli, G., Ban, Y., & Sanz, M. (2024). Satellite image classification with neural quantum kernels. arXiv preprint arXiv:2409.20356.
30. Sancho-Lorente, T., Román-Roche, J., & Zueco, D. (2022). Quantum kernels to learn the phases of quantum matter. *Physical Review A, 105*(4), 042432.
31. Wu, Y., Wu, B., Wang, J., & Yuan, X. (2023). Quantum phase recognition via quantum kernel methods. *Quantum, 7*, 981.
32. Batra, K., Zorn, K. M., Foil, D. H., Minerali, E., Gawriljuk, V. O., Lane, T. R., & Ekins, S. (2021). Quantum machine learning algorithms for drug discovery applications. *Journal of Chemical Information and Modeling, 61*(6), 2641–2647.
33. Liu, N., & Rebentrost, P. (2018). Quantum machine learning for quantum anomaly detection. *Physical Review A, 97*(4), 042315.
34. Grossi, M., Ibrahim, N., Radescu, V., Loredo, R., Voigt, K., Von Altrock, C., & Rudnik, A. (2022). Mixed quantum–classical method for fraud detection with quantum feature selection. *IEEE Transactions on Quantum Engineering, 3*, 1–12.
35. Miyabe, S., Quanz, B., Shimada, N., Mitra, A., Yamamoto, T., Rastunkov, V., Alevras, D., Metcalf, M., King, D. J. M., Mamouei, M., et al. Quantum multiple kernel learning in financial classification tasks. arXiv preprint arXiv:2312.00260.

Chapter 4
Quantum Neural Networks

Abstract This chapter presents an in-depth exploration of classical and quantum neural network paradigms, encompassing the fundamental architectures, training methodologies, and theoretical analyses of network performance. This chapter is organized into five sections: Sect. 4.1 reviews the structural and theoretical foundations of classical neural networks; Sect. 4.2 introduces the quantum perceptron model, elucidating its theoretical advantages over classical counterparts; Sect. 4.3 explores quantum neural networks (QNNs), detailing the process by which classical data is encoded into quantum states and processed through parameterized quantum gates, thereby mitigating the challenges posed by large model sizes and high computational costs; Sect. 4.4 delves into the theoretical aspects of QNNs, emphasizing their expressivity, generalization, and trainability; and finally, Sect. 4.5 provides illustrative code implementations using benchmark datasets to demonstrate the practical viability of QNNs.

Classical neural networks [1] form the bedrock of modern artificial intelligence and have achieved widespread success in domains like computer vision [2] and natural language processing [3]. However, despite these triumphs, classical neural networks grapple with significant hurdles. For instance, their excessively large model sizes and the resulting high computational demands [4] lead to substantial energy consumption [5]. These limitations stem from their reliance on classical computational resources, which become increasingly unsustainable as models grow in complexity.

Quantum neural networks (QNNs) [6] offer a promising solution by enhancing neural networks with the computational potential of quantum circuits [7]. In QNNs, classical input data is encoded into quantum states, and quantum gates with trainable parameters process these states in ways that classical systems cannot easily replicate. This computational regime leverages quantum mechanics to explore new forms of pattern recognition and problem-solving that go beyond classical methods. Thus, QNNs have the potential to outperform classical neural networks in specific learning tasks [8], where the advantages in processing and learning can be explored.

Y. Du et al., *A Gentle Introduction to Quantum Machine Learning*,
https://doi.org/10.1007/978-981-95-1284-3_4

Despite these exciting prospects, realizing the full potential of QNNs faces challenges, including quantum noise [9] and the need for scalable quantum hardware [10]. Nevertheless, continuous advancements in quantum hardware and algorithm design promise QNNs will address the inefficiencies inherent in classical models, particularly in fields such as quantum many-body physics [11] and quantum chemistry [12].

This chapter offers a systematic overview. First, the structure and function of classical neural networks are outlined in Sect. 4.1. Then, the discussion transitions to fault-tolerant and near-term quantum neural networks in Sects. 4.2 and 4.3, respectively. The theoretical foundations of QNNs are also explored in Sect. 4.4, with a focus on their **trainability** (how easily a model can "learn" from data), **expressivity** (how well they model complex patterns), and **generalization** (how well they work on new data). Finally, illustrative code implementations of QNNs using the wine [13] and MNIST datasets [14] are provided in Sect. 4.5.

4.1 Classical Neural Networks

Neural networks [15–17] are computer models inspired by the structure of the human brain, designed to process and analyze complex patterns in data. Originally developed from the concept of neurons connected by weighted pathways, neural networks have become one of the most powerful tools in artificial intelligence [1]. Each neuron processes its inputs by applying weights and nonlinear activations, producing an output that feeds into the next layer of neurons. This structure enables neural networks to learn complex functions during training [18]. For example, given a dataset of images and their labels, a neural network can learn to classify categories, such as distinguishing between cats and dogs, by adjusting its parameters during training. Guided by optimization algorithms such as gradient descent [19], the learning process allows the network to gradually reduce the error between the predicted and actual outputs, allowing it to learn the best parameters for the given task.

After nearly a century of development, neural networks have undergone remarkable advancements in both their architectures and capabilities. The simplest model, the perceptron [15], established the fundamental principle by demonstrating how neural networks could learn to separate linearly classifiable categories. Building upon this, deeper and more complex networks—such as multilayer perceptrons (MLPs) [16] and transformers [17]—have led to breakthroughs in tasks spanning autonomous driving and content generation.

4.1.1 Perceptron

The perceptron model, first introduced by McCulloch and Pitts [15], is widely recognized as a foundational structure in artificial neural networks. It has inspired architectures ranging from convolutional neural networks (CNNs) [20] and residual neural networks (ResNets) [21] to transformers [17]. Given its fundamental role, the mechanism of single-layer perceptrons is introduced next.

A single-layer perceptron comprises three fundamental components: *input neurons, a weighted layer, and an output neuron* as illustrated in Fig. 4.1. Given a d-dimensional input vector $\boldsymbol{x} \in \mathbb{R}^d$, the input layer consists of d neurons, each representing the feature $\boldsymbol{x}_i$ for $\forall i \in [d]$. This input is processed through a weighted summation, i.e.,

$$z = \boldsymbol{w}^\top \boldsymbol{x}, \tag{4.1}$$

where $\boldsymbol{w}^\top$ is the transpose of the weight vector and z is the output of the weighted layer. A nonlinear activation function is then applied to produce the output neuron $\hat{y}$. For the standard perceptron model, the sign function is typically used as the activation function:

$$\hat{y} = f(z) = \begin{cases} 1, & \text{if } z \geq 0, \\ -1, & \text{if } z < 0. \end{cases} \tag{4.2}$$

The perceptron learns from input data by iteratively adjusting its trainable parameters $\boldsymbol{w}$. In particular, let $\mathcal{D} = \left\{ \left(\boldsymbol{x}^{(a)}, y^{(a)} \right) \right\}_{a=1}^{n}$ be the training dataset, where $\boldsymbol{x}^{(a)}$ represents the input features of the a-th example, and $y^{(a)} \in \{-1, 1\}$ denotes the corresponding label. When the perceptron outputs a prediction $\hat{y}^{(s)}$, the parameters are updated accordingly, i.e.,

$$\boldsymbol{w} \leftarrow \boldsymbol{w} + \left(y^{(s)} - \hat{y}^{(s)} \right) \boldsymbol{x}^{(s)}. \tag{4.3}$$

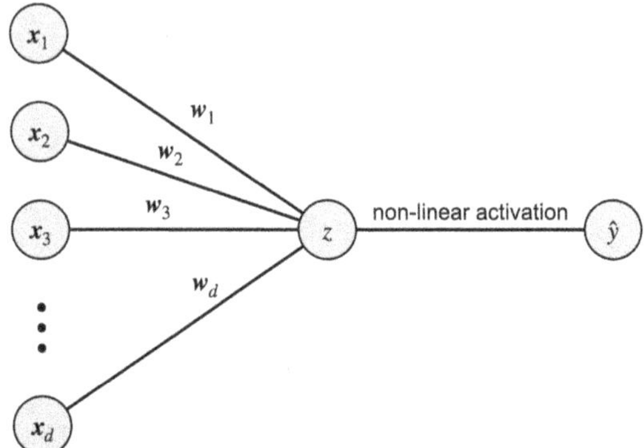

Fig. 4.1 Illustration of a perceptron

This training process repeats iteratively until the error reaches a predefined threshold.

Perceptrons can perfectly classify linearly separable data with a finite number of mistakes, as established in Theorem 4.1.

Theorem 4.1 (Convergence of Perceptrons [22]) *Suppose the training data consists of unit vectors separated by a margin of γ with labels $y^{(i)} \in \{-1, 1\}$. Then, there exists a perceptron training algorithm that achieves zero error with at most $O\left(\frac{1}{\gamma^2}\right)$ mistakes.*

Proof of Theorem 4.1 Consider the initial parameter of the perceptron $w = 0$. Since the training dataset is linearly separable by a margin of γ, there exists a unit vector w^* such that $y^{(i)} w^{*\top} x^{(i)} \geq \gamma$ for all samples $i \in [n]$. Let $x^{(s,t)}$ be the sample that is misclassified in the t-th step, which is then used for adjusting the parameter. Let $w(t)$ be the parameter after the t-th step. Using Eq. (4.3), it can be shown that

$$w^{*\top} w(t) - w^{*\top} w(t-1)$$

$$= \left(y^{(s,t)} - \hat{y}^{(s,t)} \right) w^{*\top} x^{(s,t)}$$

$$= 2 y^{(s,t)} w^{*\top} x^{(s,t)} \geq 2\gamma, \tag{4.4}$$

where Eq. (4.4) is derived by noticing the sample $\left(x^{(s,t)}, y^{(s,t)} \right)$ is misclassified with $\hat{y}^{(s,t)} \neq y^{(s,t)}$ and $y^{(s,t)}, \hat{y}^{(s,t)} \in \{-1, 1\}$. By considering the initialization $w(0) = 0$, the norm of the parameter after the t-th step can be bounded by

$$\|w(t)\| \geq \left| w^{*\top} w(t) \right| \tag{4.5}$$

$$= \left| w^{*\top} \sum_{t'=1}^{t} \left(w(t') - w(t'-1) \right) \right| \tag{4.6}$$

$$\geq 2\gamma t, \tag{4.7}$$

where Eq. (4.5) follows from the condition $\|w^*\| = 1$. Equation (4.7) is derived by using the result in Eq. (4.4). On the other hand

$$\|w(t)\|^2 - \|w(t-1)\|^2$$

$$= \left\| w(t-1) + \left(y^{(s,t)} - \hat{y}^{(s,t)} \right) x^{(s,t)} \right\|^2 - \|w(t-1)\|^2 \tag{4.8}$$

$$= \left\| w(t-1) + 2 y^{(s,t)} x^{(s,t)} \right\|^2 - \|w(t-1)\|^2 \tag{4.9}$$

$$= 4\|x^{(s,t)}\|^2 + 4 w(t-1)^\top y^{(s,t)} x^{(s,t)}$$

$$\leq 4 + 4 w(t-1)^\top y^{(s,t)} x^{(s,t)} \tag{4.10}$$

$$\leq 4, \tag{4.11}$$

where Eq. (4.8) follows from the weight update rule in Eq. (4.3). Equation (4.9) is derived by noticing that $y^{(s,t)} \neq \hat{y}^{(s,t)}$ and $y^{(s,t)}, \hat{y}^{(s,t)} \in \{-1, 1\}$. Equation (4.10) follows from the condition $\|x^{(i)}\| = 1$ for all samples. Equation (4.11) is derived by noticing that the sample $(x^{(s,t)}, y^{(s,t)})$ is misclassified by the perceptron with the parameter $w(t - 1)$, i.e.,

$$y^{(s,t)} w(t - 1)^\top x^{(s,t)} \leq 0.$$

Thus, after t steps, the parameter is bounded by

$$\|w(t)\| \leq 2\sqrt{t}. \tag{4.12}$$

Combining Eqs. (4.7) and (4.12), it can be shown that

$$t \leq \frac{1}{\gamma^2}. \tag{4.13}$$

$\square$

Since the parameters are used in an inner product operation, as shown in Eq. (4.1), the single-layer perceptron can be considered as a basic kernel method employing the identity feature mapping. Consequently, the single-layer perceptron can only classify linearly separable data and is inadequate for handling more complex tasks, such as the XOR problem [23]. This limitation has driven the development of advanced neural networks, such as multilayer perceptrons (MLPs) [16], which can capture nonlinear relationships by incorporating nonlinear activation functions and multilayer structures.

4.1.2 Multilayer Perceptron

The multilayer perceptron (MLP) is a fully connected neural network architecture consisting of three components: the *input layer, hidden layers, and output layer,* as illustrated in Fig. 4.2. Here, the dashed lines denote softmax operations. Similar to the single-layer perceptron introduced in Sect. 4.1.1, the neurons in the MLP are connected through weighted sums, followed by nonlinear activation functions.

The mathematical expression of MLP is as follows. Let $x^{(a,1)}$ be the a-th input data and $\ell = 1$ denote the input layer. Define L as the number of total layers. The forward propagation at the $(\ell + 1)$-th layer $\forall \ell \in \{1, 2, \ldots, L - 2\}$ yields

$$z^{(a,\ell+1)} = W^{(\ell)} x^{(a,\ell)} + b^{(\ell)},$$

$$x^{(a,\ell+1)} = \sigma(z^{(a,\ell+1)}),$$

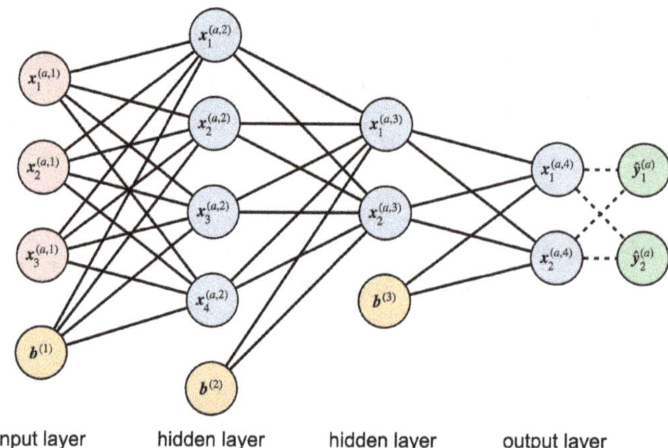

Fig. 4.2 Illustration of a multilayer perceptron with two hidden layers

Table 4.1 Common nonlinear activation functions

Name	Formulation
Sigmoid function	$\sigma(x) = 1/(1 + \exp(-x))$
Hyperbolic tangent function	$\sigma(x) = \tanh(x)$
Rectified linear unit (ReLU) function	$\sigma(x) = \max(0, x)$

where σ represents the nonlinear activation function and $W^{(\ell)}$ and $\boldsymbol{b}^{(\ell)}$ denote trainable weight and the bias term, respectively. Similar to the notation z in the perceptron in Sect. 4.1.1, $z^{(a,\ell+1)}$ denotes the output of the linear sum in the $\ell + 1$-th layer, which is expressed in a more generalized vector form. Therefore, the parameters for the weighted linear sum are expressed in matrix form as $W^{(\ell)}$. Various methods exist for implementing nonlinear activations, with some common approaches summarized in Table 4.1.

After passing through $L - 2$ hidden layers, the output of MLP given by the equation below serves as the prediction to approximate the target label $\boldsymbol{y}^{(a)}$, i.e.,

$$\hat{\boldsymbol{y}}^{(a)} = \mathrm{softmax}\left(\boldsymbol{x}^{(a,L)}\right) := \frac{\left(\exp\left(\boldsymbol{x}_1^{(a,L)}\right), \cdots, \exp\left(\boldsymbol{x}_p^{(a,L)}\right)\right)^{\top}}{\sum_{i=1}^{p} \exp\left(\boldsymbol{x}_i^{(a,L)}\right)},$$

with p here denotes the dimension of $\boldsymbol{x}^{(a,L)}$.

Next, consider a simple binary classification example to illustrate the MLP learning process. Let $\{(\boldsymbol{x}^{(a)}, \boldsymbol{y}^{(a)})\}_{a \in \mathcal{D}}$ be the training dataset $\mathcal{D}$, where $\boldsymbol{x}^{(a)}$ is the feature vector and $\boldsymbol{y}^{(a)} \in \{(1, 0)^{\top}, (0, 1)^{\top}\}$ is the label for two categories. Consider the MLP with one hidden layer. The prediction can be expressed as follows:

$$\hat{y}^{(a)} = \text{softmax}(x^{(a,3)}) = \text{softmax} \circ \sigma \left(z^{(a,3)} \right)$$

$$= \text{softmax} \circ \sigma \left(W^{(2)} x^{(a,2)} + b^{(2)} \right)$$

$$= \text{softmax} \circ \sigma \left(W^{(2)} \sigma \left(z^{(a,2)} \right) + b^{(2)} \right)$$

$$= \text{softmax} \circ \sigma \left(W^{(2)} \sigma \left(W^{(1)} x^{(a,1)} + b^{(1)} \right) + b^{(2)} \right),$$

where $\circ$ denotes the function composition. Here, $\sigma(x) = 1/(1 + \exp(-x))$ is used as the nonlinear activation function.

MLP learns from the given dataset by minimizing the loss function with respect to the parameters $\theta = (W^{(1)}, W^{(2)}, b^{(1)}, b^{(2)})$. A possible choice of the loss function is the ℓ_2 norm distance between the prediction and the label:

$$\mathcal{L}(\theta) = \frac{1}{|\mathcal{D}|} \sum_{a \in \mathcal{D}} \mathcal{L}^{(a)}(\theta) = \frac{1}{2|\mathcal{D}|} \sum_{a \in \mathcal{D}} \left\| \hat{y}^{(a)}(\theta) - y^{(a)} \right\|^2. \tag{4.14}$$

We use gradient descent with learning rate η to optimize the parameters:

$$\theta(t + 1) = \theta(t) - \eta \nabla_\theta \mathcal{L}(\theta(t)).$$

As illustrated in Fig. 4.3, the gradient is computed using backpropagation [24] as follows. First, the gradient with respect to the output layer is given by

$$\frac{\partial \mathcal{L}^{(a)}}{\partial \hat{y}^{(a)}} = \hat{y}^{(a)} - y^{(a)},$$

$$\frac{\partial \mathcal{L}^{(a)}}{\partial x^{(a,3)}} = \frac{\partial \hat{y}^{(a)}}{\partial x^{(a,3)}} \frac{\partial \mathcal{L}^{(a)}}{\partial \hat{y}^{(a)}} = \left[\text{diag} \left(\hat{y}^{(a)} \right) - \hat{y}^{(a)} \hat{y}^{(a)\top} \right] \frac{\partial \mathcal{L}^{(a)}}{\partial \hat{y}^{(a)}},$$

$$\frac{\partial \mathcal{L}^{(a)}}{\partial z^{(a,3)}} = \frac{\partial x^{(a,3)}}{\partial z^{(a,3)}} \frac{\partial \mathcal{L}^{(a)}}{\partial x^{(a,3)}} = \text{diag} \left[\left(1 - z^{(a,3)} \right) \odot z^{(a,3)} \right] \frac{\partial \mathcal{L}^{(a)}}{\partial x^{(a,3)}},$$

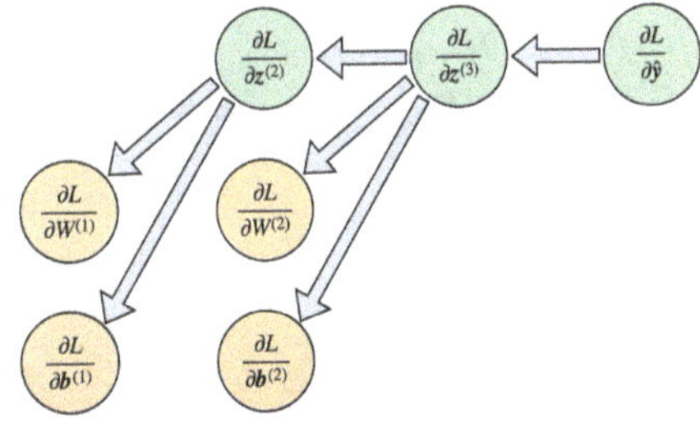

Fig. 4.3 Illustration of backpropagation when calculating the gradient of an MLP with one hidden layer. The index of sample a is omitted for simplicity

where $\mathbf{1}$ denotes the vector $(1, 1, \cdots, 1)^\top$ and $\odot$ denotes the element-wise multiplication (Hadamard product). For convenience, we omit the dimension of $\mathbf{1}$ here, which has the same dimension with $z^{(a,3)}$. Next, the gradient with respect to the hidden layer can be obtained using the chain rule:

$$\frac{\partial \mathcal{L}^{(a)}}{\partial x^{(a,2)}} = \frac{\partial z^{(a,3)}}{\partial x^{(a,2)}} \frac{\partial \mathcal{L}^{(a)}}{\partial z^{(a,3)}} = W^{(2)} \frac{\partial \mathcal{L}^{(a)}}{\partial z^{(a,3)}},$$

$$\frac{\partial \mathcal{L}^{(a)}}{\partial W^{(2)}} = \frac{\partial z^{(a,3)}}{\partial W^{(2)}} \frac{\partial \mathcal{L}^{(a)}}{\partial z^{(a,3)}} = \frac{\partial \mathcal{L}^{(a)}}{\partial z^{(a,3)}} x^{(a,2)\top},$$

$$\frac{\partial \mathcal{L}^{(a)}}{\partial b^{(2)}} = \frac{\partial z^{(a,3)}}{\partial b^{(2)}} \frac{\partial \mathcal{L}^{(a)}}{\partial z^{(a,3)}} = \frac{\partial \mathcal{L}^{(a)}}{\partial z^{(a,3)}},$$

$$\frac{\partial \mathcal{L}^{(a)}}{\partial z^{(a,2)}} = \frac{\partial x^{(a,2)}}{\partial z^{(a,2)}} \frac{\partial \mathcal{L}^{(a)}}{\partial x^{(a,2)}} = \mathrm{diag}\left[\left(\mathbf{1} - z^{(a,2)}\right) \odot z^{(a,2)}\right] \frac{\partial \mathcal{L}^{(a)}}{\partial x^{(a,2)}}.$$

The gradient with respect to the parameters for the input layer is derived similarly:

$$\frac{\partial \mathcal{L}^{(a)}}{\partial W^{(1)}} = \frac{\partial z^{(a,2)}}{\partial W^{(1)}} \frac{\partial \mathcal{L}^{(a)}}{\partial z^{(a,2)}} = \frac{\partial \mathcal{L}^{(a)}}{\partial z^{(a,2)}} x^{(a,1)\top},$$

$$\frac{\partial \mathcal{L}^{(a)}}{\partial b^{(1)}} = \frac{\partial z^{(a,2)}}{\partial b^{(1)}} \frac{\partial \mathcal{L}^{(a)}}{\partial z^{(a,2)}} = \frac{\partial \mathcal{L}^{(a)}}{\partial z^{(a,2)}}.$$

After multiple training epochs, the loss function converges to a value below a predefined threshold, which leads to a small classification error.

Compared to single-layer perceptrons, MLPs can model nonlinear relationships by employing hidden layers and activation functions. This enables them to learn abstract representations by capturing the complex patterns inherent in the data. Mathematically, the power of MLPs is guaranteed by the universal approximation theorem, as stated in Theorem 4.1, which asserts that a single hidden layer is sufficient to approximate any arbitrary continuous function.

Fact 4.1 (Universal Approximation Theorem, Informal Version Adapted from [25]) *Let $C(\mathcal{X}, \mathbb{R}^m)$ denote the set of continuous functions from a subset X of a Euclidean space $\mathbb{R}^n$ to a Euclidean space $\mathbb{R}^m$. Denote by σ a function that is not polynomial. Then, for every $n, m \in \mathbb{N}$, compact set $\mathcal{K} \subseteq \mathbb{R}^n$, $f \in C(\mathcal{K}, \mathbb{R}^m)$, and $\epsilon > 0$, there exist $k \in \mathbb{N}$, $A \in \mathbb{R}^{k \times n}$, $b \in \mathbb{R}^k$, and $C \in \mathbb{R}^{m \times k}$ such that*

$$\sup_{x \in X} \| f(x) - g(x) \| < \epsilon,$$

where $g(x) = C\sigma(Ax + b)$.

> **Remark**
> MLPs involve a large number of parameters due to their fully connected
> multilayer architecture. This high parameter count provides MLPs with
> considerable flexibility (how well they model complex patterns), allowing
> them to learn complex data distributions. However, the excessive capacity to
> fit the training data often leads to overfitting [26], where the MLP captures
> noise and irrelevant patterns instead of generalizable features. As a result,
> MLPs tend to perform poorly on unseen data, especially when the training
> set is limited or noisy. To mitigate this issue, advanced techniques such
> as dropout [27], weight decay [28], and attention mechanisms [17] have
> been proposed to reduce overfitting in MLPs while maintaining sufficient
> expressivity.

4.2 Fault-Tolerant Quantum Perceptron

The primary goal of advancing quantum machine learning (QML) is to leverage the
computational power of quantum mechanics to improve performance across various
learning tasks. As discussed in Sect. 1.1.2, these advantages can include reduced
runtime, lower query complexity, and enhanced sample efficiency compared to
classical models. A notable example is the quantum perceptron model [29]. This
FTQC-based QML algorithm, which uses the Grover search, offers a quadratic
improvement in the query complexity during training over its classical counterpart.
For a complete understanding, this section first introduces the Grover search
algorithm, followed by a detailed explanation of the quantum perceptron model.

4.2.1 Grover Search

Grover search [30] provides runtime speedups for unstructured search problems,
finding broad use in cryptography, quantum machine learning, and constraint
satisfaction. Classical search methods typically require $O(d)$ queries for a dataset
with d entries. In contrast, Grover's algorithm can identify the target element with
high probability using only $O\left(\sqrt{d}\right)$ queries to a quantum oracle. Consequently,
quantum algorithms that incorporate Grover search have the potential to achieve a
quadratic speedup over classical approaches.

In general, a search task can be abstracted as a function $f(x)$ such that $f(x) = 1$
if x belongs to the solution set of the search problem and $f(x) = 0$ otherwise. We
consider a dataset consisting of $d = 2^N$ elements, where each element is represented
by the quantum state $|x\rangle$ with $x = 0, 1, \cdots, d-1$. In this process, two key quantum

oracles are introduced. The first oracle, $U_0 = 2(|0\rangle\langle 0|)^{\otimes N} - \mathbb{I}_d$, applies a phase shift of $e^{i\pi} = -1$ to all quantum states except $|0\rangle^{\otimes N}$, which remains unchanged. The second oracle, U_f, operates in a similar manner. That is, it applies a phase shift of -1 to quantum states that belong to the solution set while leaving all other states unaffected. The procedure for Grover search is described in Algorithm 2.

Algorithm 2 Grover search

Require: Quantum oracles U_f and U_0. The size of the dataset and the solution set, denoted by $d = 2^N$ and M, respectively.
Ensure: An index corresponds to one of the solution states with high probability.
 1: Initialize a register of N qubits with the state of uniform superposition:

$$|\phi_0\rangle = \frac{1}{\sqrt{d}} \sum_{x=0}^{d-1} |x\rangle = \bigotimes_{n=1}^{N} \frac{|0\rangle + |1\rangle}{\sqrt{2}} = \left(\bigotimes_{n=1}^{N} \mathrm{H} \right) |0\rangle.$$

 2: Let $m = \lfloor \frac{\pi}{4} \sqrt{\frac{d}{M}} - \frac{1}{2} \rfloor$. Apply the following operation:

$$|\phi_m\rangle = \left[\mathrm{H}^{\otimes N} U_0 \mathrm{H}^{\otimes N} U_f \right]^m |\phi_0\rangle.$$

 3: Measure the state $|\phi_m\rangle$ to generate an index.

Theorem 4.2 (Time Complexity of Grover Search) *Grover search finds a solution to the unstructured search problem with high probability in time* $O\left(\sqrt{\frac{d}{M}} (\log d + T_f) \right)$, *where d is the size of the dataset, M is the size of the solution set, and T_f denotes the time complexity of implementing the oracle U_f.*

Remark
The Grover search achieves a quadratic speedup in the query complexity of the oracle U_f. It provides a quantum advantage in the runtime only if the time complexity of the oracle U_f, denoted as T_f, is less than $\sqrt{d}$.

Proof of Theorem 4.2 Let the superposition of the solution state be

$$|\text{target}\rangle = \frac{1}{\sqrt{M}} \sum_{x:f(x)=1} |x\rangle.$$

Similarly, the superposition of the states outside the solution set is given by

$$|\text{other}\rangle = \frac{1}{\sqrt{2^N - M}} \sum_{x|f(x)=0} |x\rangle.$$

Thus, the initial uniform superposition state can be expressed as

$$|\phi_0\rangle = \sqrt{\frac{M}{2^N}}|\text{target}\rangle + \sqrt{\frac{2^N - M}{2^N}}|\text{other}\rangle := \alpha_0|\text{target}\rangle + \beta_0|\text{other}\rangle.$$

In principle, the coefficient associated with the target state is expected to increase during the quantum state evolution, such that the solution could be obtained through quantum measurement with high probability. The dynamics of these coefficients can be described as follows:

$$\begin{aligned}
\alpha_k &= \langle\text{target}|\phi_k\rangle \\
&= \langle\text{target}|H^{\otimes N}U_0 H^{\otimes N}U_f|\phi_{k-1}\rangle \\
&= \langle\text{target}|\,(2|\phi_0\rangle\langle\phi_0| - \mathbb{I})\,U_f\,(\alpha_{k-1}|\text{target}\rangle + \beta_{k-1}|\text{other}\rangle) \\
&= \langle\text{target}|\,(2|\phi_0\rangle\langle\phi_0| - \mathbb{I})\,(-\alpha_{k-1}|\text{target}\rangle + \beta_{k-1}|\text{other}\rangle) \\
&= \left(1 - 2\alpha_0^2\right)\alpha_{k-1} + 2\alpha_0\beta_0\beta_{k-1}, \\
\beta_k &= \langle\text{other}|\phi_k\rangle \\
&= \langle\text{other}|\,(2|\phi_0\rangle\langle\phi_0| - \mathbb{I})\,(-\alpha_{k-1}|\text{target}\rangle + \beta_{k-1}|\text{other}\rangle) \\
&= \left(2\beta_0^2 - 1\right)\beta_{k-1} - 2\alpha_0\beta_0\alpha_{k-1}.
\end{aligned}$$

Let the angle $\theta = \arccos\sqrt{\frac{2^N - M}{2^N}}$, then by induction, it can be shown that

$$\alpha_k = \sin[(2k+1)\theta], \quad \beta_k = \cos[(2k+1)\theta].$$

To ensure that the coefficient $\alpha_m = O(1)$, there is a condition $(2m+1)\theta \approx \pi/2$. Therefore, $m = O(1/\theta) = O\left(\sqrt{\frac{2^N}{M}}\right)$ suffices to obtain the solution with high probability. $\qquad\square$

4.2.2   Online Quantum Perceptron with Quadratic Speedups

As stated in Theorem 4.1, for a linearly separable dataset with a margin γ, a perceptron model can achieve perfect classification after making $O(1/\gamma^2)$ mistakes during training. In classical approaches, identifying a sample that is misclassified by the current model may require up to $O(d)$ queries, where d denotes the size of the training dataset. In contrast, the quantum perceptron model [29] can find misclassified samples more efficiently using the Grover search algorithm, achieving a quadratic speedup in the number of queries needed.

First, let's look at how the input data is set up. For classification, a dataset $\{z^{(i)}\}_{i=1}^{d} = \{(x^{(i)}, y^{(i)})\}_{i=1}^{d}$ is considered, where the label $y^{(i)} \in \{-1, 1\}$. For convenience, it is assumed that the number of samples, d, is a power of 2, i.e., $d = 2^N$. Each data vector $x^{(i)}$ is represented using B bits. The information of each sample $z^{(i)}$ is stored in the quantum state $|z^{(i)}\rangle$ using $B + 1$ qubits.

Example 4.1 For the sample $(x^{(i)}, y^{(i)}) = ([0, 0, 1, 0], 1)$, the corresponding quantum state is $|z^{(i)}\rangle = |00101\rangle$. Here, the last qubit encodes the label (where "0" represents the label "-1"), and the other qubits represent the data vector. If $x^{(i)}$ is a float vector, a similar bit sequence can be can be formed by concatenating the binary representations of its elements.

Next, the oracle models are introduced. A quantum oracle U is assumed to exist for encoding training data as the corresponding quantum state, i.e.,

$$U|i\rangle|0\rangle = |i\rangle|z^{(i)}\rangle \,, \quad U^{\dagger}|i\rangle|z^{(i)}\rangle = |i\rangle|0\rangle. \tag{4.15}$$

Due to the linearity of unitary operations

$$U \sum_{i=0}^{d-1} \frac{1}{\sqrt{d}}|i\rangle|0\rangle = \sum_{i=0}^{d-1} \frac{1}{\sqrt{d}}|i\rangle|z^{(i)}\rangle. \tag{4.16}$$

In addition to the input oracle U described in Eq. (4.15), the quantum perceptron model employs another oracle to distinguish between correctly classified and misclassified quantum states. Specifically, the oracle F'_{w} satisfies

$$F'_{w}|z^{(i)}\rangle = (-1)^{f(w, z^{(i)})}|z^{(i)}\rangle, \tag{4.17}$$

where $f : (w, z^{(i)}) \rightarrow \{0, 1\}$. The function outputs 1 if the current perceptron model with weight w misclassifies the training sample $z^{(j)}$; otherwise, it outputs 0. Furthermore, we define

$$F_{w} = U^{\dagger}(\mathbb{I} \otimes F'_{w})U, \tag{4.18}$$

which is used as the oracle U_f in the Grover search. The online quantum perceptron procedure is given in Algorithm 3. The query complexity of the online quantum perceptron is provided in Theorem 4.3.

Algorithm 3 Online quantum perceptron

Require: Linearly separable dataset $\{z^{(i)}\}_{i=1}^{d} = \{(x^{(i)}, y^{(i)})\}_{i=1}^{d}$, where $d = 2^N$. Margin threshold γ. Constants $\epsilon \in (0, 1)$ and $c \in (1, 2)$.
Ensure: Weight w for a perceptron that correctly classifies the dataset with a margin γ with probability at least $1 - \epsilon$.
1: Initialize the weight $w = 0$.
2: **for** $h = 1, \cdots, \lceil \frac{1}{\gamma^2} \rceil$ **do**
3:     **for** $k = 1, \cdots, \lceil \log_{3/4} \gamma^2 \epsilon \rceil$ **do**
4:         **for** $j = 1, \cdots, \lceil \log_c \frac{1}{\sin(2 \sin^{-1}(1/\sqrt{d}))} \rceil$ **do**
5:             Draw m uniformly from $\{0, \cdots, \lceil c^j \rceil - 1\}$.
6:             Prepare the quantum state

$$|\phi_0\rangle = \frac{1}{\sqrt{d}} \sum_{i=0}^{d-1} |i\rangle.$$

7:             Generate the state

$$|\phi_1\rangle = \{[(2|\phi_0\rangle\langle\phi_0| - \mathbb{I}_d) \otimes \mathbb{I}_d] F_w\}^m |\phi_0\rangle |0\rangle^{\otimes N}.$$

8:             Measure the first register of the state $|\phi_1\rangle$ to obtain an outcome q.
9:             **if** $f(w, z^{(q)}) = 1$ **then**
10:                 Update $w \leftarrow w + y^{(q)} x^{(q)}$.
11:             **end if**
12:         **end for**
13:     **end for**
14: **end for**
15: Output w.

Theorem 4.3 (Online Quantum Perceptron [29]) *Consider a training dataset that consists of unit vectors $\{x^{(1)}, \cdots, x^{(d)}\}$ and labels $\{y^{(1)}, \cdots, y^{(d)}\}$ with a margin γ. Denote by n_{quant} the number of queries to F_w needed to learn the weight w, such that the training dataset is perfectly classified with probability at least $1 - \epsilon$, then*

$$n_{\text{quant}} \in O\left(\frac{\sqrt{d}}{\gamma^2} \log \frac{1}{\gamma^2 \epsilon}\right).$$

For the classical case where the training vectors are uniformly sampled from the training dataset, the number of queries to f_w is bounded by

$$\Omega(d) \ni n_{\text{class}} \in O\left(\frac{d}{\gamma^2} \log \frac{1}{\gamma^2 \epsilon}\right).$$

Proof of Theorem 4.3 The main idea of the quantum perceptron model in Algorithm 3 is to replace the procedure of finding the misclassified sample in classical perceptrons with the Grover search. Due to convergence result for perceptrons in Theorem 4.1, $h = 1, \cdots, \lceil \frac{1}{\gamma^2} \rceil$ iterations of Steps (3–13) suffice to update the weight w toward the case of perfect classification. Therefore, Theorem 4.3 is the direct consequence of the following lemmas and Theorem 4.1. The query complexity of classical perceptrons has the lower bound $\Omega(d)$, since the model needs to go through the entire dataset in the worst case. $\qquad\square$

Lemma 4.1 *Given only uniform sampling access to the training dataset, there exists a classical perceptron that either finds a misclassified sample to update the weight w or concludes that no such example exists with probability $1 - \epsilon\gamma^2$, using $O(d \log(1/\epsilon\gamma^2))$ queries to f_w.*

Lemma 4.2 *The procedure of Steps 3–13 in Algorithm 3 either finds a misclassified sample to update the weight w or concludes that no such example exists with probability $1 - \epsilon\gamma^2$, using $O\left(\sqrt{d} \log(1/\epsilon\gamma^2)\right)$ queries to F_w.*

Proof (Proof of Lemma 4.1) First, let $m_c = d\lceil \log(1/\epsilon\gamma^2) \rceil$ be the number of samples drawn from the dataset uniformly in each iteration of training. Suppose these samples are classified correctly, then the probability that the entire dataset is classified correctly is

$$\text{Pr(Correct classification)} \geq 1 - \left(1 - \frac{1}{d}\right)^{m_c} \geq 1 - \exp\left(-\frac{m_c}{d}\right) \geq 1 - \epsilon\gamma^2.$$

$\qquad\square$

Proof of Lemma 4.2 For convenience, denote $\theta_a := \arccos\sqrt{\frac{d-d_0}{d}}$, where d_0 the number of misclassified samples in the dataset according to the current model. Let $d_1 := \lceil \log_c \frac{1}{\sin(2\sin^{-1}(1/\sqrt{d}))} \rceil$. Here, an exponential expansion strategy is used in Steps 4–12 to handle the scenario of unknown d_0. Namely, quantum operations in the Grover search are repeated for m times, where m is drawn from an exponentially expanded set $0, \cdots, \lceil c^j \rceil - 1$ uniformly for a predefined $c \in (1, 2)$ and $j = 1, \cdots, d_1$. It can be shown that this strategy can find a misclassified sample before the convergence of Algorithm 3 with an average probability at least $1/4$:

$$\Pr\left(f(\boldsymbol{w}, z^{(q)}) = 1\right) = \sum_{j=1}^{d_1} \frac{1}{\lceil c^j \rceil} \sum_{m=0}^{\lceil c^j \rceil - 1} \sin^2((2m+1)\theta_a)$$

$$\geq \frac{1}{\lceil c^{d_1} \rceil} \sum_{m=0}^{\lceil c^{d_1} \rceil - 1} \sin^2((2m+1)\theta_a)$$

$$= \frac{1}{2}\left[1 - \frac{\sin(4\lceil c^{d_1} \rceil \theta_a)}{2\lceil c^{d_1} \rceil \sin(2\theta_a)}\right]$$

$$\geq \frac{1}{4}.$$

The procedure of Steps 4–12 is repeated for $k = 1, \cdots, \lceil \log_{3/4} \gamma^2 \epsilon \rceil$ iterations to accumulate the success probability. The probability of finding a misclassified sample in Steps 3–13 before the convergence of Algorithm 3 is at least

$$1 - \left(1 - \frac{1}{4}\right)^{\lceil \log_{3/4} \epsilon \gamma^2 \rceil} \geq 1 - \epsilon \gamma^2. \tag{4.19}$$

Finally, the query complexity Q of Steps 3–13 in Algorithm 3 can be upper bounded as follows:

$$Q \leq \sum_{k=1}^{\lceil \log_{3/4} \gamma^2 \epsilon \rceil} \sum_{j=1}^{d_1} c^j$$

$$\leq \left(1 + \log_{3/4} \gamma^2 \epsilon\right) \frac{c}{1-c}\left[1 - c^{d_1}\right]$$

$$\leq \left(1 + \log_{3/4} \gamma^2 \epsilon\right) \frac{c^2}{c-1}\left[\frac{1}{\sin(2\sin^{-1}(1/\sqrt{d}))} - 1\right]$$

$$= O\left(\sqrt{d} \log \frac{1}{\epsilon \gamma^2}\right).$$

$\square$

4.3 NISQ-Era Quantum Neural Networks

Following recent experimental breakthroughs in superconducting quantum hardware [10, 31–33], researchers have devoted considerable effort to developing and implementing quantum machine learning algorithms optimized for current and near-term quantum devices [34]. Unlike fault-tolerant quantum computers, these devices face three primary limitations: quantum noise, limited coherence time, and circuit connectivity constraints. Regarding quantum noise, state-of-the-art devices have single-qubit gate error rates of $10^{-4} \sim 10^{-3}$ and two-qubit gate error rates of approximately $10^{-3} \sim 10^{-2}$ [32, 33]. The coherence time is around $10^2 \, \mu$s [10, 32, 33], primarily limited by decoherence in noisy quantum channels. Regarding circuit connectivity, most superconducting quantum processors employ architectures that exhibit two-dimensional connectivity patterns and their variants [10, 32, 33]. Gate operations between nonadjacent qubits must be executed through intermediate relay operations, leading to additional error accumulation. To address these inherent limitations, the quantum neural network (QNN) framework has been proposed. Specifically, these QNNs are designed to perform meaningful computations on near-term quantum devices.

4.3.1 General Framework

This section introduces the basic architecture of QNNs. As illustrated in Fig. 4.4, a fundamental QNN comprises three main parts: the input, the model circuit, and the measurement. Specifically, the input state ρ_{in} is prepared using the operation U_{data}, followed by a variational quantum circuit (VQC) $V(\boldsymbol{\theta})$ and the measurement operation.

Input The QNN uses quantum states ρ_{in} as input data. As shown in Table 4.2, QNNs can process both classical and quantum data. Specifically, the input states ρ_{in} may be introduced from physical processes such as quantum Hamiltonian evolutions. Another way is to construct the state ρ_{in} by encoding classical vectors using encoding protocols introduced in Sect. 2.3.1, such as angle encoding and amplitude encoding.

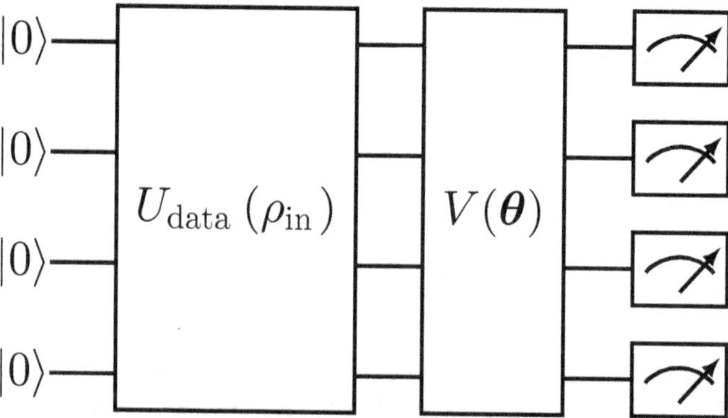

Fig. 4.4 Illustration of a QNN

Table 4.2 Examples of classical and quantum data employed in QNNs, where H denotes the system Hamiltonian, k_B is the Boltzmann constant, and $|\phi_0\rangle$ is a predefined initial state

	Example	Input state formulation
Classical data	Angle encoding	$\bigotimes_{n=1}^{N} [\mathrm{RY}(x_n)|0\rangle]$
	Amplitude encoding	$\sum_{i=0}^{d-1} x_i / \|x\|_2 |i\rangle$
Quantum data	Gibbs state	$\dfrac{\exp(-H/k_B T)}{\mathrm{Tr}[\exp(-H/k_B T)]}$
	Hamiltonian evolution	$\exp(-iHt)|\phi_0\rangle$

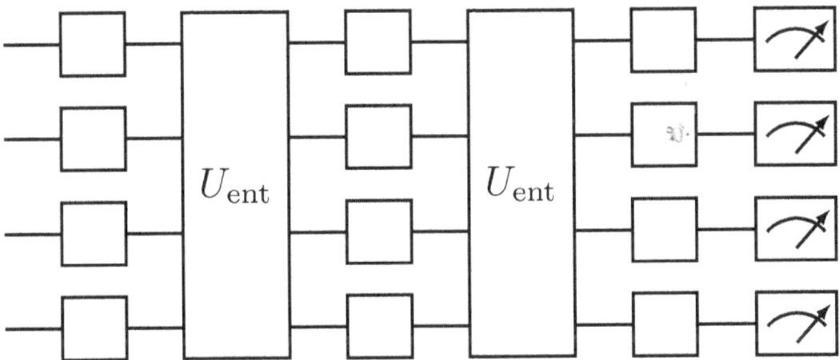

Fig. 4.5 Illustration of a hardware-efficient circuit with two entanglement layers

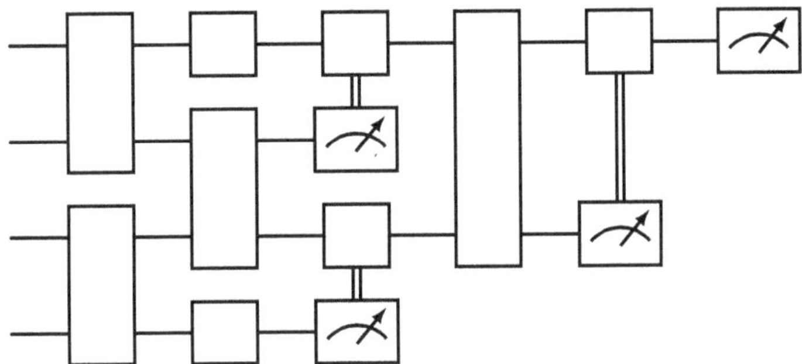

Fig. 4.6 Illustration of a quantum convolutional neural network

Model Circuit QNNs employ variational quantum circuits (VQCs), a.k.a, ansatzes, to extract and learn features from input data. A typical VQC, denoted as $V(\theta)$, features a layered structure composed of both parameterized (trainable) and fixed quantum gates. For general-purpose implementations, a common strategy is to use the parameters θ as the phases of single-qubit rotation gates RX, RY, RZ, while quantum entanglement is introduced through fixed two-qubit gates, such as CX and CZ. Standard circuit architectures include the hardware-efficient circuit (HEC) [35], shown in Fig. 4.5, and the quantum convolutional neural network (QCNN) [36], shown in Fig. 4.6. For problem-specific applications, such as finding the ground states of molecular Hamiltonians, specialized circuits like the unitary coupled cluster *ansatz* [37] are employed.

Example 4.2 Hardware-efficient circuits incorporate several widely adopted *ansatzes*. Single-qubit rotations {RX, RY, RZ} are used to construct parameterized single-qubit unitaries. The entangled unitary layer can be implemented using two-qubit gates such as

$$U_{\text{ent}} = \bigotimes_{n=1}^{\lceil \frac{N}{2} \rceil} U(2n - 1 + k\%2, 2n + k\%2) \quad \text{for the } k\text{-th layer,}$$

$$U_{\text{ent}} = \bigotimes_{n=1}^{\lceil \frac{N}{2} \rceil} U(2n - 1, 2n) \bigotimes_{n=1}^{\lceil \frac{N-1}{2} \rceil} U(2n, 2n + 1),$$

where $U \in \{CX, CZ\}$.

Measurement After implementing the model circuit, the quantum state is measured using specific observables, denoted as O, to extract classical information. The choice of observables depends on the experimental objectives. For a variational quantum eigensolver, where the goal is to find the ground state and energy of a given Hamiltonian, the observable is chosen to be the target Hamiltonian itself. In quantum machine learning applications involving classical data, the measurement outcomes are used to approximate label information, which typically lacks direct physical significance. As a result, the observable can, in principle, be any Hermitian operator. However, for practical experimental considerations, a linear combination of Pauli-Z operators is commonly used as the observable:

$$O = \sum_{j=1}^{N} c_j \mathbb{I}^{\otimes(j-1)} \otimes Z_j \otimes \mathbb{I}^{\otimes(N-j)}, \tag{4.20}$$

where $c \in \mathbb{R}^N$ is a weight vector. The measurement outcome of QNN can be expressed as a function of θ, i.e.,

$$f(\theta; \rho_{\text{in}}, V, O) = \text{Tr}\left[O V(\theta) \rho_{\text{in}} V(\theta)^\dagger \right]. \tag{4.21}$$

Training of QNNs As a QML framework, the optimization of QNNs amounts to updating parameters θ using gradient-based methods. Thanks to the linearity of quantum mechanics and the unitary evolution constraint, in certain cases, gradients can be elegantly calculated using the parameter-shift rule.

Theorem 4.4 (Parameter-Shift Rule [38]) *Suppose the gate $G_j(\theta_j)$ in a VQC $V(\theta)$ has a unitary Hamiltonian H_j, then the corresponding gradient could be obtained as*

$$\frac{\partial f}{\partial \theta_j}(\theta) = \frac{1}{2}\left[f\left(\theta + \frac{\pi}{2}e^{(j)}\right) - f\left(\theta - \frac{\pi}{2}e^{(j)}\right)\right],$$

where the function f follows the Eq. (4.21) and the one-hot vector $e^{(j)}$ has the same dimension with θ with the j-th element being 1.

Proof (Proof of Theorem 4.4)
  For convenience, we denote the detailed structure of VQC as

$$V(\theta) = \prod_{i=L}^{1} G_i(\theta_i)W_i,$$

where L is the number of parameters in VQC, G_i is the parameterized gate, and W_i is the fixed gate. By assumption, the gate takes the form as

$$G_j(\theta_j) = \exp(-i H_j \theta_j/2),$$

where the Hamiltonian H_j is a unitary. For convenience, unnecessary parameterized and fixed gates can be merged into the state ρ_{in} and the observable O, i.e.,

$$\rho'_{\text{in}} = W_j \left(\prod_{i=j-1}^{1} G_i(\theta_i)W_i\right) \rho_{\text{in}} \left(\prod_{i=1}^{j-1} W_i^\dagger G_i(\theta_i)^\dagger\right) W_j^\dagger,$$

$$O' = \left(\prod_{i=j}^{L} W_i^\dagger G_i(\theta_i)^\dagger\right) O \left(\prod_{i=L}^{j} G_i(\theta_i)W_i\right).$$

It can be shown that

$$\begin{aligned}
f(\theta) &= \text{Tr}\left[O V(\theta)\rho_{\text{in}} V(\theta)^\dagger\right]\\
&= \text{Tr}\left[O' G_j(\theta_j)\rho'_{\text{in}} G_j(\theta_j)^\dagger\right]\\
&= \text{Tr}\left[O' \exp(-i H_j \theta_j/2)\rho'_{\text{in}} \exp(i H_j \theta_j/2)\right]\\
&= \cos^2 \frac{\theta_j}{2} \text{Tr}\left[O'\rho'_{\text{in}}\right] + \frac{i}{2}\sin \theta_j \left[[H_j, O']\rho'_{\text{in}}\right] + \sin^2 \frac{\theta_j}{2} \text{Tr}\left[H_j O' H_j \rho'_{\text{in}}\right],
\end{aligned}$$

$$(4.22)$$

where $[A, B] := AB - BA$ denotes the commutator.

After some calculations from Eq. (4.22), it can be shown that

$$f\left(\boldsymbol{\theta} + \frac{\pi}{2}\boldsymbol{e}^{(j)}\right) = \frac{1-\sin\theta_j}{2}\,\mathrm{Tr}\left[O'\rho'_{\mathrm{in}}\right] + \frac{i}{2}\cos\theta_j\left[[H_j, O']\rho'_{\mathrm{in}}\right]$$

$$+\frac{1+\sin\theta_j}{2}\,\mathrm{Tr}\left[H_j O' H_j \rho'_{\mathrm{in}}\right]$$

$$f\left(\boldsymbol{\theta} - \frac{\pi}{2}\boldsymbol{e}^{(j)}\right) = \frac{1+\sin\theta_j}{2}\,\mathrm{Tr}\left[O'\rho'_{\mathrm{in}}\right] - \frac{i}{2}\cos\theta_j\left[[H_j, O']\rho'_{\mathrm{in}}\right]$$

$$+\frac{1-\sin\theta_j}{2}\,\mathrm{Tr}\left[H_j O' H_j \rho'_{\mathrm{in}}\right]$$

$$\frac{\partial f}{\partial\theta_j}(\boldsymbol{\theta}) = -\frac{1}{2}\sin\theta_j\,\mathrm{Tr}\left[O'\rho'_{\mathrm{in}}\right] + \frac{i}{2}\cos\theta_j\left[[H_j, O']\rho'_{\mathrm{in}}\right]$$

$$+\frac{1}{2}\sin\theta_j\,\mathrm{Tr}\left[H_j O' H_j \rho'_{\mathrm{in}}\right].$$

Comparing the above equations, Theorem 4.4 is proved. $\square$

4.3.2   Discriminative Learning with QNNs

This section presents an example of using a QNN for discriminative learning, which focuses on distinguishing between different categories. The focus is on binary classification, where the label $y^{(i)} = \pm 1$ corresponds to the input state $\rho^{(i)}$. For classical data, the state $\rho^{(i)} = |\psi(\boldsymbol{x}^{(i)})\rangle\langle\psi(\boldsymbol{x}^{(i)})|$ can be generated from the classical vector $\boldsymbol{x}^{(i)}$ using a read-in approach:

$$|\psi(\boldsymbol{x}^{(i)})\rangle = U_\phi(\boldsymbol{x}^{(i)})|0\rangle. \tag{4.23}$$

Here, the feature map can be constructed via angle encoding, as introduced in Sect. 2.3.1:

$$U_\phi(\boldsymbol{x}^{(i)}) = \bigotimes_{n=1}^{N} \mathrm{RY}(x_n^{(i)}) = \bigotimes_{n=1}^{N} \exp(-iYx_n^{(i)}/2). \tag{4.24}$$

Denote by O and $V(\boldsymbol{\theta})$ the quantum observable and the VQC, respectively. The prediction function of the QNN is given by

$$\hat{y}^{(i)}(\boldsymbol{\theta}) = \mathrm{Tr}[O V(\boldsymbol{\theta})\rho^{(i)} V(\boldsymbol{\theta})^\dagger]. \tag{4.25}$$

In the binary classification task, the QNN learns by training the parameter θ to minimize the distance between the label $y^{(i)}$ and the prediction $\hat{y}^{(i)}(\theta)$. Specifically, the mean square error (MSE) is used as the loss function:

$$\theta^* = \mathrm{argmin}\,\mathcal{L}(\theta), \quad \text{where } \mathcal{L}(\theta) = \sum_{i=1}^{n} \ell(\theta, x^{(i)}, y^{(i)}) = \frac{1}{2} \sum_{i=1}^{n} \left(\hat{y}^{(i)}(\theta) - y^{(i)} \right)^2. \tag{4.26}$$

The gradient of the loss in Eq. (4.26) can be calculated via the chain rule, i.e.,

$$\nabla_\theta \mathcal{L}(\theta) = \sum_{i=1}^{n} \left(\hat{y}^{(i)}(\theta) - y^{(i)} \right) \nabla_\theta \hat{y}^{(i)}(\theta), \tag{4.27}$$

where the gradient of the prediction $\hat{y}^{(i)}$ can be obtained by using the parameter-shift rule in Theorem 4.4. Consequently, a variety of gradient-based optimization algorithms, such as stochastic gradient descent [19], Adagrad [39], and Adam [40], can be employed to train QNNs.

> **Remark**
> The QNN binary classification framework can be naturally extended to multi-label classification using the **one-vs-all** strategy. Specifically, we train k QNN binary classifiers for k classes, with each classifier distinguishing a specific class from the others.

> **Remark**
> The QNN classification framework presented in this section can be extended to quantum regression learning by incorporating continuous labels.

4.3.3 Generative Learning with QNNs

This section introduces a quantum generative model implemented by QNNs: the quantum generative adversarial network (QGAN) [41]. Like its classical counterparts, QGAN learns to generate samples by using a discriminator and a generator that compete in a two-player minimax game. Specifically, both the discriminator D and the generator G can be implemented using QNNs. By leveraging the expressive power of QNNs, QGAN has the potential to exhibit quantum advantages in certain tasks [42, 43].

To illustrate QGAN's training and sampling processes, two examples based on the quantum patch and batch GANs, as proposed by Huang et al. [44], are presented. Let N denote the number of qubits and M the number of training samples. The patch and batch strategies are designed for the cases where $N < \lceil \log M \rceil$ and $N > \lceil \log M \rceil$, respectively. In particular, the patch strategy allows for generating high-dimensional images even with limited quantum resources. In contrast, the batch strategy supports parallel training when enough quantum resources are accessible.

4.3.3.1 Quantum Patch Generative Adversarial Network

The discussion begins with the quantum patch GAN, which consists of a quantum generator, as illustrated in Fig. 4.7, a classical discriminator, and a classical optimizer. Both the learning and sampling processes of an image are performed in patches, involving T sub-generators. For the t-th sub-generator, the model takes a latent state z as input and generates a sample $G_t(z)$. Specifically, the latent state is prepared from the initial state $|0\rangle^{\otimes N}$ using a single-qubit rotation layer, where the parameters $\{\alpha_n\}_{n=1}^{N}$ are sampled from the uniform distribution over $[0, 2\pi)$. The latent state is then processed through an N-qubit hardware-efficient circuit $U_{G_t}(\boldsymbol{\theta})$, which leads to the state

$$|\psi_t(z)\rangle = U_{G_t}(\boldsymbol{\theta})|z\rangle. \tag{4.28}$$

To perform nonlinear operations, partial measurements are conducted, and a subsystem $\mathcal{A}$ (ancillary qubits) is traced out from the state $|\psi_t(z)\rangle$. The resulting

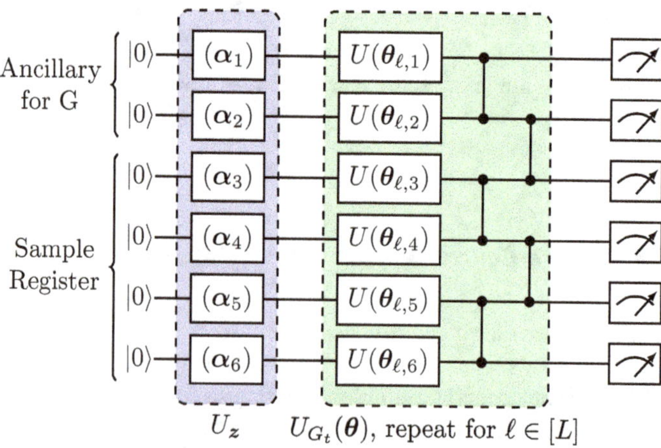

Fig. 4.7 The quantum generator used in the quantum patch GAN, where each $U(\boldsymbol{\theta}_{\ell,n}) \in \mathcal{U}(2)$ is a trainable single-qubit unitary

mixed state is

$$\rho_t(z) = \frac{\mathrm{Tr}_{\mathcal{A}}\left[\Pi \otimes \mathbb{I} |\psi_t(z)\rangle\langle\psi_t(z)|\right]}{\mathrm{Tr}\left[\Pi \otimes \mathbb{I} |\psi_t(z)\rangle\langle\psi_t(z)|\right]}, \tag{4.29}$$

where Π is the projective operator acting on the subsystem $\mathcal{A}$. Subsequently, the mixed state $\rho_t(z)$ is measured in the computational basis to obtain the sample $G_t(z)$. Specifically, let $\Pr(J = j) := \mathrm{Tr}[|j\rangle\langle j|\rho_t(z)]$, where the probabilities of the outcomes can be estimated by the measurement. The sample $G_t(z)$ is then defined as

$$G_t(z) = [\Pr(J = 0), \cdots, \Pr(J = j), \cdots, \Pr(J = 2^{N-N_{\mathcal{A}}} - 1)], \tag{4.30}$$

where $N_{\mathcal{A}}$ is the number of qubits in $\mathcal{A}$. Finally, the complete image is reconstructed by aggregating these samples from all sub-generators as follows:

$$G(z) = [G_1(z), \cdots, G_T(z)]. \tag{4.31}$$

In principle, the discriminator D in a quantum patch GAN can be any classical neural network that takes the training data x or the generated sample $G(z)$ as input, with the output

$$D(x),\ D(G(z)) \in [0, 1]. \tag{4.32}$$

Let γ and θ denote the parameters of the discriminator D and the generator G, respectively. The optimization problem for the quantum patch GAN can be formulated as

$$\min_{\theta} \max_{\gamma} \mathcal{L}(D_\gamma(G_\theta(z)), D_\gamma(x)) := \underset{x}{\mathbb{E}}\left[\log D_\gamma(x)\right] + \underset{z}{\mathbb{E}}\left[\log(1 - D_\gamma(G_\theta(z)))\right]. \tag{4.33}$$

Similar to quantum discriminative learning, the quantum patch GAN can be trained using gradient-based optimization algorithms.

4.3.3.2 Quantum Batch GAN

As illustrated in Fig. 4.8, the quantum batch GAN differs from the quantum patch GAN by employing a quantum discriminator. In a quantum batch GAN, all qubits are divided into two registers: the index register, consisting of N_I qubits, and the feature register, consisting of N_F qubits. The qubits in the feature register are further partitioned into three parts: N_D qubits for generating quantum samples, N_{A_G} qubits for implementing nonlinear operations in the generator G_θ, and N_{A_D} qubits for implementing nonlinear operations in the discriminator D_γ. For a batch with size

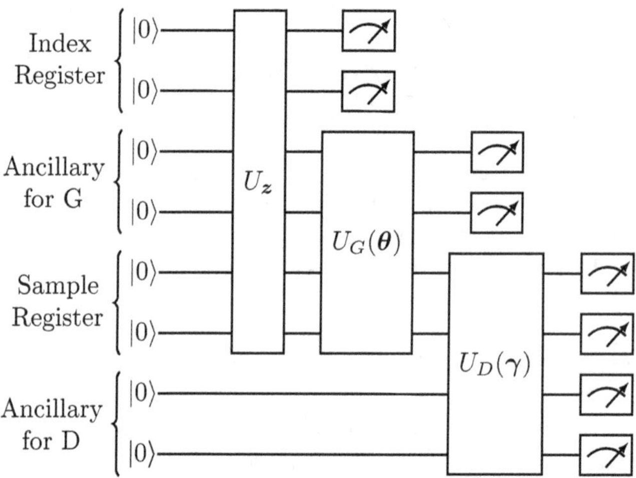

Fig. 4.8 The main structure of the quantum batch GAN

$|B_k| = 2^{N_I}$, two oracles are used to encode the information of latent vectors and training samples:

$$|0\rangle_I \otimes |0\rangle_F \xrightarrow{U_z} \frac{1}{2^{N_I}} \sum_i |i\rangle_I \otimes |z^{(i)}\rangle_F, \tag{4.34}$$

$$|0\rangle_I \otimes |0\rangle_F \xrightarrow{U_x} \frac{1}{2^{N_I}} \sum_i |i\rangle_I \otimes |x^{(i)}\rangle_F. \tag{4.35}$$

Remark

For data with M features, state preparation for amplitude encoding in U_x requires $\tilde{O}(2^{N_I} M)$ multi-controlled quantum gates, which is infeasible for current quantum devices. This challenge can be addressed by employing pretrained shallow circuit approximations of the given oracle [45].

After the encoding stage, a PQC $U_G(\theta)$ and the corresponding partial measurement are employed as the quantum generator. Thus, the generated state corresponding to $|B_k|$ fake samples is obtained as follows:

$$\frac{1}{2^{N_I}} \sum_i |i\rangle_I \otimes |z^{(i)}\rangle_F$$

$$\xrightarrow{U_G(\boldsymbol{\theta})} \frac{1}{2^{N_I}} \sum_i |i\rangle_I \otimes \left(U_G(\boldsymbol{\theta}) \otimes \mathbb{I}_{2^{N_{A_D}}} |z^{(i)}\rangle_F\right) := |\psi(z)\rangle$$

$$\xrightarrow{\Pi_{A_G}} \frac{\mathbb{I}_{2^{N_I}} \otimes \Pi_{A_G} \otimes \mathbb{I}_{2^{N_D+N_{A_D}}} |\psi(z)\rangle}{\mathrm{Tr}\left[\mathbb{I}_{2^{N_I}} \otimes \Pi_{A_G} \otimes \mathbb{I}_{2^{N_D+N_{A_D}}} |\psi(z)\rangle\langle\psi(z)|\right]} := |G_{\boldsymbol{\theta}}(z)\rangle,$$

where the partial measurement $\Pi_{A_G} = (|0\rangle\langle0|)^{\otimes N_{A_G}}$ serves as the nonlinear operation. In the sampling stage, the reconstructed image is generated similarly to the quantum patch GAN. Specifically, the i-th image $G_{\boldsymbol{\theta}}(z^{(i)})$ in the batch is

$$G_{\boldsymbol{\theta}}(z^{(i)}) = \left[\mathrm{Pr}(J = 0|I = i), \cdots, \mathrm{Pr}(J = 2^{N_D} - 1|I = i)\right], \tag{4.36}$$

where

$$\mathrm{Pr}(J = j|I = i) = \mathrm{Tr}\left[|i\rangle_I|j\rangle_F \langle i|_I \langle j|_F |G(z)\rangle\langle G(z)|\right]. \tag{4.37}$$

Finally, we introduce the training stage. A quantum discriminator is applied to either the fake generated state $|G_{\boldsymbol{\theta}}(z)\rangle$ or the real data state $|x\rangle$. Similar to the quantum generator, the quantum discriminator $D_{\boldsymbol{\gamma}}$ consists of a PQC $U_D(\boldsymbol{\gamma})$, followed by the corresponding partial measurement. In the case of the real state, the state evolution proceeds as follows:

$$\frac{1}{2^{N_I}} \sum_i |i\rangle_I \otimes |x^{(i)}\rangle_F$$

$$\xrightarrow{U_D(\boldsymbol{\gamma})} \frac{1}{2^{N_I}} \sum_i |i\rangle_I \otimes \left(\mathbb{I}_{2^{N_{A_G}}} \otimes U_D(\boldsymbol{\gamma})|x^{(i)}\rangle_F\right) := |\psi(x)\rangle$$

$$\xrightarrow{\Pi_{A_D}} \frac{\mathbb{I}_{2^{N-N_{A_D}}} \otimes \Pi_{A_G}|\psi(x)\rangle}{\mathrm{Tr}\left[\mathbb{I}_{2^{N-N_{A_D}}} \otimes \Pi_{A_G}|\psi(x)\rangle\langle\psi(x)|\right]} := |D_{\boldsymbol{\gamma}}(x)\rangle,$$

where the partial measurement is $\Pi_{A_G} = (|0\rangle\langle0|)^{\otimes N_{A_D}}$. The classical description $D_{\boldsymbol{\gamma}}(x)$ is generated similarly to Eq. (4.36). The generated state $G_{\boldsymbol{\theta}}|z\rangle$ undergoes the same procedure to obtain the description $D_{\boldsymbol{\gamma}}(G_{\boldsymbol{\theta}}(z))$. These classical vectors are then used in the loss function in Eq. (4.33) to train parameters $\boldsymbol{\theta}$ and $\boldsymbol{\gamma}$.

4.4 Theoretical Foundations of Quantum Neural Networks

Quantum neural networks (QNNs) primarily aim to make accurate predictions on unseen data. Achieving this goal depends on three key factors: expressivity, generalization ability, and trainability, as illustrated in Fig. 4.9. The expressivity of a QNN defines its hypothesis space $\mathcal{H}$ (solid blue ellipse). When $\mathcal{H}$ has a moderate size and encompasses the target concept (solid red star), QNNs can achieve good performance. Conversely, if $\mathcal{H}$ is too small to cover the target concept (the solid gray star), the QNN's performance diminishes. During QNN optimization, a significant challenge is the vanishing gradient problem, often called the barren plateau. This issue prohibits a good estimation near the target parameters θ^*. A thorough analysis of these factors is crucial for understanding QNNs' potential advantages and limitations relative to classical machine learning models. Instead of providing an exhaustive review of all theoretical results, this section highlights key conceptual insights of QNNs.

As explained in Sect. 3.3, expressivity refers to a model's ability to represent a wide range of functions, determining the smallest achievable training error. Section 4.4.1 characterizes the expressivity of QNNs using the *covering number*, an advanced tool from statistical learning theory. This analysis will reveal the relationship between the expressivity of QNNs and their structural factors, such as the size of the quantum system and the number of exploited quantum gates. Understanding this connection helps clarify how QNNs' expressivity scales with their architecture.

Generalization ability evaluates the discrepancy between a model's performance on the training data and on unseen test data. Section 4.4.1 further explores the relationship between the generalization ability and expressivity of QNNs by deriving a generalization error bound in terms of the covering number. This bound provides insights into how the expressivity of QNNs—specifically their structural factors—may impact their ability to generalize and offers a framework to assess their potential advantages over classical ML models.

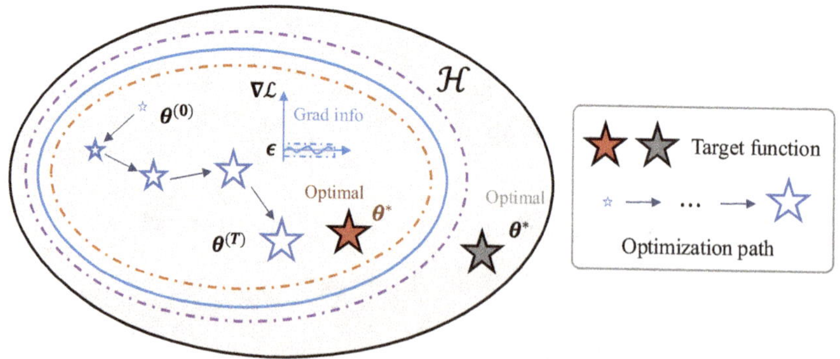

Fig. 4.9 Overview of the expressivity, generalization ability, and trainability of QNNs

While expressivity and generalization ability are crucial for both quantum kernels and QNNs, trainability emerges as an additional consideration for QNNs due to the introduction of trainable parameters in quantum circuits. This leads to fundamentally different optimization challenges, rendering many existing results from classical ML models inapplicable. Specifically, trainability refers to a model's ability to efficiently converge to a good solution during training, directly influencing the computational cost. In Sect. 4.4.2, the barren plateau problem is introduced. This is a major challenge in training quantum neural networks. As the system size increases, gradients vanish exponentially. This makes optimization extremely difficult. Additionally, various strategies to address this issue will be discussed, offering practical insights into enhancing the trainability of QNNs.

4.4.1 Expressivity and Generalization of Quantum Neural Networks

The expressivity and generalization are deeply interconnected within the framework of statistical learning theory for understanding the prediction ability of any learning model. To better understand these terms in the context of quantum neural networks, let us first review the framework of empirical risk minimization (ERM), which is a popular framework for analyzing these abilities in statistical learning theory.

Let $\mathcal{D} = \{(\boldsymbol{x}^{(i)}, y^{(i)})\}_{i=1}^n \in \mathcal{X} \times \mathcal{Y}$ be the training dataset sampled independently from an unknown distribution $\mathcal{P}$. A learning algorithm $\mathcal{A}$ aims to use the dataset $\mathcal{D}$ to infer a hypothesis $h_{\theta^*} : \mathcal{X} \to \mathcal{Y}$ from the hypothesis space $\mathcal{H}$ that could accurately predict all labels of $\boldsymbol{x} \in \mathcal{X}$ following the distribution $\mathcal{P}$. This amounts to identifying an optimal hypothesis in $\mathcal{H}$ minimizing the expected risk:

$$R(h) = \mathbb{E}_{(\boldsymbol{x},y)\sim\mathcal{P}}\ell(h_{\theta^*}(\boldsymbol{x}), y), \tag{4.38}$$

where $\ell(\cdot, \cdot)$ refers to the per-sample loss predefined by the learner. Unfortunately, since the distribution $\mathcal{P}$ is unknown, the expected risk cannot be directly assessed. In practice, $\mathcal{A}$ alternatively learns an empirical hypothesis $h_{\hat{\theta}} \in \mathcal{H}$, as the global minimizer of the (regularized) loss function

$$\mathcal{L}(\boldsymbol{\theta}, \mathcal{D}) = \frac{1}{n}\sum_{i=1}^n \ell(h_{\boldsymbol{\theta}}(\boldsymbol{x}^{(i)}), y^{(i)}) + \mathcal{R}(\boldsymbol{\theta}), \tag{4.39}$$

where $\mathcal{R}(\boldsymbol{\theta})$ refers to an optional regularizer, as will be detailed in the following. Moreover, the first term on the right-hand side refers to the empirical risk:

$$R_{\mathrm{ERM}}(h_{\hat{\theta}}) = \frac{1}{n}\sum_{i=1}^n \ell(h_{\hat{\theta}}(\boldsymbol{x}^{(i)}), y^{(i)}), \tag{4.40}$$

which is also known as the training error. To address the intractability of $R(h_{\hat{\theta}})$, one can decompose it into two measurable terms:

$$R(h_{\hat{\theta}}) = R_{\mathrm{ERM}}(h_{\hat{\theta}}) + R_{\mathrm{Gene}}(h_{\hat{\theta}}), \tag{4.41}$$

where $R_{\mathrm{Gene}}(h_{\hat{\theta}}) = R(h_{\hat{\theta}}) - R_{\mathrm{ERM}}(h_{\hat{\theta}})$ refers to the generalization error. In this regard, a small prediction error necessitates the learning model to achieve both a small training error and a small generalization error.

4.4.1.1 An Overview

Before moving to analyze the training error (ERM) and generalization error of QNNs rigorously, we first delve into better understanding the meaning of expressivity and generalization ability of the learning models with the ERM framework. Moreover, an intuition about the necessities and benefits of exploring such theoretical aspects of QNNs as a special learning model is provided.

Expressivity can be directly understood as the size of the learning model's hypothesis space $\mathcal{H} = \{h_{\theta} : \theta \in \Theta\}$. Intuitively, the achievable smallest empirical risk is determined by the expressivity of learning models. Specifically, a learning model with low expressivity may not fit the training data with complex patterns, e.g., the hypothesis space of linear model $\mathcal{H} = \{h_{\theta} = \theta \cdot x\}$ cannot fit the nonlinear data $\{x^{(i)}, (x^{(i)})^2\}$ perfectly.

In general, the cardinality of the hypothesis space is infinity, as the parameters θ are continuous. This makes it hard to compare the expressivity of different learning models. An alternative measure is model complexity, which measures the richness of the hypothesis space through the structural factors of the specific learning models, such as the number of parameters, depth, or architectural design. Remarkably, model complexity is measurable and bounded. In this tutorial, the covering number will be used to measure the model complexity of QNNs.

The generalization capability of learning models is directly measured by the generalization error R_{Gene} in Eq. (4.41). Good generalization means the learning model predicts well on both unseen and training data. In this regard, a small generalization error combined with a small training error implies a small prediction error, as the generalization error ensures that prediction performance on unseen data is comparable to training performance.

In statistical learning theory, it is well established that a bias-variance trade-off governs the interplay between model complexity and generalization performance for any learning model. This highlights the delicate balance required for a model to generalize well to unseen data. The relationship is often depicted by a U-shaped curve, as shown in Fig. 4.10. This curve suggests that there exists an optimal level of model complexity for improving the generalization ability of any learning model. When under the point related to optimal expressivity, increasing model complexity improves performance on training data and enhances generalization. However, beyond a certain point, higher complexity leads to overfitting, resulting

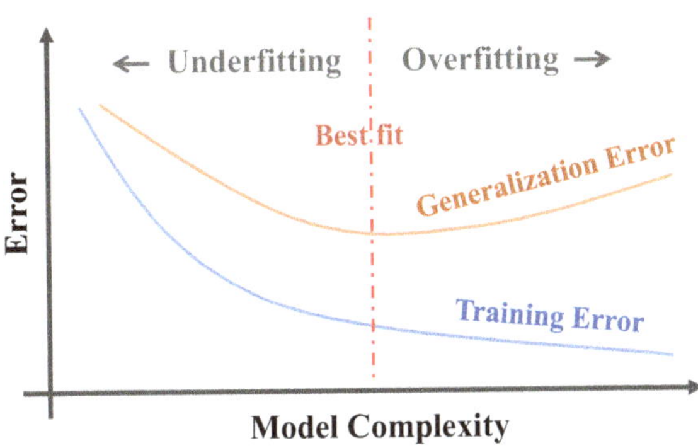

Fig. 4.10 Influence of model complexity on generalization error

in poor generalization on test data. For QNNs, identifying this optimal level of complexity is crucial for achieving the best balance between training performance and generalization.

4.4.1.2 Expressivity of QNNs

In this chapter, we analyze the generalization error of QNNs through a specific measure of model complexity: the covering number. This measure helps us better understand and characterize the generalization performance of QNNs.

To elucidate the specific definition of the covering number, the general structures of QNNs are first reviewed. Define $\rho \in \mathbb{C}^{2^N \times 2^N}$ as the N-qubit input quantum states, $O \in \mathbb{C}^{2^N \times 2^N}$ as the quantum observable, $U(\boldsymbol{\theta}) = \prod_{l=1}^{N_g} u_l(\boldsymbol{\theta}) \in \mathcal{U}(2^N)$ as the applied ansatz, where $\boldsymbol{\theta} \in \Theta$ are the trainable parameters living in the parameter space Θ, $u_l(\boldsymbol{\theta}) \in \mathcal{U}(2^k)$ refers to the l-th quantum gate operated with at most k-qubits with $k \leq N$, and $\mathcal{U}(2^N)$ stands for the unitary group in dimension 2^N. In general, $U(\boldsymbol{\theta})$ is formed by N_{gt} trainable gates and $N_g - N_{gt}$ fixed gates, e.g., $\Theta \subset [0, 2\pi)^{N_{gt}}$. Under the above definitions, the explicit form of the output of QNN under the ideal scenarios is

$$h(\boldsymbol{\theta}, O, \rho) := \mathrm{Tr}\left(U(\boldsymbol{\theta})^\dagger O U(\boldsymbol{\theta})\rho\right). \tag{4.42}$$

Given the training dataset $\mathcal{D} = \{(\rho^{(i)}, y^{(i)})\}_{i=1}^n$ and loss function $\mathcal{L}(\boldsymbol{\theta}, \mathcal{D})$ defined in Eq. (4.39), QNN is optimized to find a good approximation $h^*(\boldsymbol{\theta}, O, \rho) = \arg\min_{h(\boldsymbol{\theta}, O, \rho) \in \mathcal{H}} \mathcal{L}(\boldsymbol{\theta}, \mathcal{D})$ that can well approximate the target concept, where $\mathcal{H}$ refers to the hypothesis space of QNNs with

$$\mathcal{H} = \left\{\mathrm{Tr}\left(U(\boldsymbol{\theta})^\dagger O U(\boldsymbol{\theta})\rho\right) \Big| \boldsymbol{\theta} \in \Theta\right\}. \tag{4.43}$$

An intuition about how the hypothesis space $\mathcal{H}$ affects the performance of QNNs is depicted in Fig. 4.9. When $\mathcal{H}$ has a modest size and covers the target concepts,

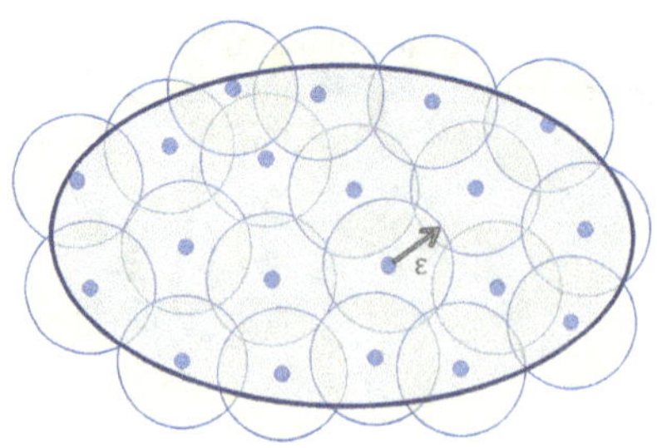

Fig. 4.11 The geometric intuition of covering number. Covering number concerns the minimum number of spherical balls with radius ϵ that occupy the whole space

the estimated hypothesis could well approximate the target concept. By contrast, when the complexity of $\mathcal{H}$ is too low, a significant gap exists between the estimated hypothesis and the target concept. An effective measure to evaluate the complexity of $\mathcal{H}$ is covering number, an advanced tool broadly used in statistical learning theory, to bound the complexity of $\mathcal{H}$ and measure the expressivity of QNNs.

Definition 4.1 (Covering Number) The covering number $\mathcal{N}(\mathcal{U}, \epsilon, \|\cdot\|)$ denotes the least cardinality of any subset $V \subset U$ that covers U at scale ϵ with a norm $\|\cdot\|$, i.e., $\sup_{A \in \mathcal{U}} \min_{B \in V} \|A - B\| \leq \epsilon$. Here, we use this notion to measure the expressivity of QNNs.

The geometric interpretation of the covering number is depicted in Fig. 4.11, which refers to the minimum number of spherical balls with radius ϵ that are required to completely cover a given space with possible overlaps. This notion has been employed to study other crucial topics in quantum physics such as Hamiltonian simulation and entangled states. Note that ϵ is a predefined hyperparameter, i.e., a small constant with $\epsilon \in (0, 1)$, and is independent of any factor. This convention is widely adopted in machine learning to evaluate the capacity of various models.

Following the convention of [46], a step-by-step analysis of the model complexity of the hypothesis space $\mathcal{H}$ of QNNs defined in Eq. (4.43) is provided. In particular, how the covering number of QNNs is controlled by their structural factors will be elucidated, including the number of parameterized gates N_{gt}, the number of qubits k the gates acting on, and the type of the quantum observable O. To this end, we first consider a simpler hypothesis space consisting of the operator group:

$$\mathcal{H}_{\text{circ}} := \left\{ U(\boldsymbol{\theta})^{\dagger} O U(\boldsymbol{\theta}) \middle| \boldsymbol{\theta} \in \Theta \right\}. \tag{4.44}$$

This space removes the input state ρ is removed compared to the hypothesis space $\mathcal{H}$ related to QNNs. Actually, the covering number of $\mathcal{H}$ under the metric d could be connected to the covering number of $\mathcal{H}_{\text{circ}}$ under the related metric d_{circ} through employing their Lipschitz properties, which is encapsulated in the following fact.

Fact 4.2 *Let $(\mathcal{H}_1, d_1)$ and $(\mathcal{H}_2, d_2)$ be two metric spaces satisfying $f : \mathcal{H}_1 \to \mathcal{H}_2$ be bi-Lipschitz such that*

$$c_l d_1(\boldsymbol{x}, \boldsymbol{z}) \leq d_2(f(\boldsymbol{x}), f(\boldsymbol{z})) \leq c_r d_1(\boldsymbol{x}, \boldsymbol{z}), \quad \forall \boldsymbol{x}, \boldsymbol{z} \in \mathcal{H}_1. \tag{4.45}$$

Then, their covering number obeys

$$\mathcal{N}(\mathcal{H}_1, 2\epsilon/c_l, d_1) \leq \mathcal{N}(\mathcal{H}_2, \epsilon, d_2) \leq \mathcal{N}(\mathcal{H}_1, \epsilon/c_r, d_1), \tag{4.46}$$

where the left inequality requires $\epsilon \leq c_l c_u/2$ with c_u being the upper bound of the distance between any two points in $\mathcal{H}_1$, namely, $d_1(x, z) \leq c_u$ for $x, z \in \mathcal{H}_1$.

Fact 4.2 indicates that we can derive the covering number of the metric space $(\mathcal{H}, d)$ by analyzing the covering number of the metric space $(\mathcal{H}_{\text{circ}}, d_{\text{circ}})$ and the Lipschitz constants of the mapping between $\mathcal{H}$ and $\mathcal{H}_{\text{circ}}$. Intuitively, a quantum circuit with many multi-qubit parameterized gates results in a complex QNN with high model complexity. These intuitions are formalized into Theorem 4.5. Specifically, the result of the covering number of the metric space $(\mathcal{H}_{\text{circ}}, d_{\text{circ}})$ is encapsulated in Lemma 4.3.

Lemma 4.3 *Suppose that the employed N-qubit quantum circuit containing in total N_g gates with $N_g > N$, each gate $u_i(\boldsymbol{\theta})$ acting on most k qubits, and $N_{gt} \leq N_g$ gates in $U(\boldsymbol{\theta})$ are trainable. The ϵ-covering number for the operator group $\mathcal{H}_{\text{circ}}$ in Eq. (4.44) with respect to the operator-norm distance obeys*

$$\mathcal{N}(\mathcal{H}_{\text{circ}}, \epsilon, \|\cdot\|) \leq \left(\frac{7N_{gt}\|O\|}{\epsilon}\right)^{2^{2k}N_{gt}}, \tag{4.47}$$

where $\|O\|$ denotes the operator norm of O.

Proof of Lemma 4.3 To measure the covering number the operator group of $\mathcal{H}_{\text{circ}} = \{U(\boldsymbol{\theta})^{\dagger} O U(\boldsymbol{\theta}) | \boldsymbol{\theta} \in \Theta\}$, one could first consider a fixed ϵ-covering $\mathcal{S}$ for the set $\mathcal{N}(U(2^k), \epsilon, \|\cdot\|)$ of all possible gates and define the set

$$\tilde{\mathcal{S}} := \left\{ \prod_{i \in \{N_{gt}\}} u_i(\boldsymbol{\theta}_i) \prod_{j \in \{N_g - N_{gt}\}} u_j \,\middle|\, u_i(\boldsymbol{\theta}_i) \in \mathcal{S} \right\}, \tag{4.48}$$

where $u_i(\boldsymbol{\theta}_i)$ and u_j specify the trainable and fixed quantum gates in the employed quantum circuit, respectively. Note that for any circuit $U(\boldsymbol{\theta}) = \prod_{i=1}^{N_g} u_i(\boldsymbol{\theta}_i)$, one can always find a $U_\epsilon(\boldsymbol{\theta}) \in \tilde{\mathcal{S}}$ where each $u_i(\boldsymbol{\theta}_i)$ of trainable gates is replaced with the nearest element in *the covering set $\mathcal{S}$*, and the discrepancy $\|U(\boldsymbol{\theta})^{\dagger} O U(\boldsymbol{\theta}) - U_\epsilon(\boldsymbol{\theta})^{\dagger} O U_\epsilon(\boldsymbol{\theta})\|$ satisfies

$$\|U(\boldsymbol{\theta})^{\dagger} O U(\boldsymbol{\theta}) - U_\epsilon(\boldsymbol{\theta})^{\dagger} O U_\epsilon(\boldsymbol{\theta})\|$$

$$\leq \|U - U_\epsilon\|\|O\|$$

$$\leq N_{gt}\|O\|\epsilon, \tag{4.49}$$

where the first inequality uses the triangle inequality and the second inequality follows from $\|U - U_\epsilon\| \leq N_{gt}\epsilon$.

Therefore, by Definition 4.1, $\tilde{S}$ forms an $N_{gt}\|O\|\epsilon$-covering set for $\mathcal{H}_{\text{circ}}$. An upper bound for the group S, as established by Barthel and Lu [47, Lemma 1], gives $|S| \leq \left(\frac{7}{\epsilon}\right)^{2^{2k}}$. Since there are $|S|^{N_{gt}}$ combinations for the gates in $\tilde{S}$, it follows that $|\tilde{S}| \leq \left(\frac{7}{\epsilon}\right)^{2^{2k} N_{gt}}$ and the covering number for $\mathcal{H}_{\text{circ}}$ satisfies

$$N(\mathcal{H}_{\text{circ}}, N_{gt}\|O\|\epsilon, \|\cdot\|) \leq \left(\frac{7}{\epsilon}\right)^{2^{2k} N_{gt}}. \tag{4.50}$$

An equivalent representation of the above inequality is

$$N(\mathcal{H}_{\text{circ}}, \epsilon, \|\cdot\|) \leq \left(\frac{7N_{gt}\|O\|}{\epsilon}\right)^{2^{2k} N_{gt}}. \tag{4.51}$$

$\square$

With the established covering number of operator group $\mathcal{H}_{\text{circ}}$, one could directly analyze the covering number of the hypothesis space $\mathcal{H}$ related to QNNs, which is encapsulated in the following theorem.

Theorem 4.5 *For $0 < \epsilon < 1/10$, the covering number of the hypothesis space $\mathcal{H}$ in Eq. (4.43) yields*

$$N(\mathcal{H}, \epsilon, |\cdot|) \leq \left(\frac{7N_{gt}\|O\|}{\epsilon}\right)^{2^{2k} N_{gt}}, \tag{4.52}$$

where $\|O\|$ denotes the operator norm of O.

Proof of Theorem 4.5 The intuition of the proof is as follows. Recall the definition of the hypothesis space $\mathcal{H}$ in Eq. (4.43) and Lemma 4.2. When $\mathcal{H}_1$ refers to the hypothesis space $\mathcal{H}$ and $\mathcal{H}_2$ refers to the unitary group $\mathcal{U}(2^N)$, the upper bound of the covering number of $\mathcal{H}$, i.e., $N(\mathcal{H}_1, d_1, \epsilon)$, can be derived by first quantifying c_r Eq. (4.45) and then interacting with $N(\mathcal{H}_{\text{circ}}, \epsilon, \|\cdot\|)$ in Lemma 4.3. Based on the above observations, the following addresses the upper bound of the covering number $N(\mathcal{H}, \epsilon, |\cdot|)$.

The Lipschitz constant c_r in Eq. (4.45) is derived as a prerequisite for establishing the upper bound of $N(\mathcal{H}, \epsilon, |\cdot|)$. Define $U \in \mathcal{U}(2^N)$ as the employed quantum circuit composed of N_g gates, i.e., $U = \prod_{i=1}^{N_g} u_l$. Let U_ϵ be the quantum circuit where each of the N_g gates is replaced by the nearest element in the covering set. The relation between the distance $d_2(\text{Tr}(U_\epsilon^\dagger O U_\epsilon \rho), \text{Tr}(U^\dagger O U \rho))$ and the distance $d_1(U_\epsilon, U)$ yields

$$d_2(\mathrm{Tr}(U_\epsilon^\dagger OU_\epsilon\rho), \mathrm{Tr}(U^\dagger OU\rho))$$

$$=|\mathrm{Tr}(U_\epsilon^\dagger OU_\epsilon\rho) - \mathrm{Tr}(U^\dagger OU\rho)|$$

$$=\left|\mathrm{Tr}\left((U_\epsilon^\dagger OU_\epsilon - U^\dagger OU)\rho\right)\right|$$

$$\leq \left\|U_\epsilon^\dagger OU_\epsilon - U^\dagger OU\right\| \mathrm{Tr}(\rho)$$

$$=d_1(U_\epsilon^\dagger OU_\epsilon, U^\dagger OU), \tag{4.53}$$

where the first equality comes from the explicit form of the hypothesis, the first inequality uses the Cauchy-Schwartz inequality, and the last inequality employs $\mathrm{Tr}(\rho) = 1$ and

$$\left\|U_\epsilon^\dagger OU_\epsilon - U^\dagger OU\right\| = d_1(U_\epsilon^\dagger OU_\epsilon, U^\dagger OU). \tag{4.54}$$

The above equation indicates $c_r = 1$. Combining the above result with Lemma 4.2 (i.e., Eq. (4.45)) and Lemma 4.3, we obtain

$$\mathcal{N}(\mathcal{H}, \epsilon, |\cdot|) \leq \mathcal{N}(\mathcal{H}_{\mathrm{circ}}, \epsilon, \|\cdot\|) \leq \left(\frac{7N_{gt}\|O\|}{\epsilon}\right)^{2^{2k}N_{gt}}. \tag{4.55}$$

This relation ensures

$$\mathcal{N}(\mathcal{H}, \epsilon, |\cdot|) \leq \left(\frac{7N_{gt}\|O\|}{\epsilon}\right)^{2^{2k}N_{gt}}. \tag{4.56}$$

$$\square$$

Theorem 4.5 indicates that the most decisive factor, which controls the complexity of $\mathcal{H}$, is the employed quantum gates in $U(\boldsymbol{\theta})$. This claim is ensured by the fact that the term $2^{2^k N_{gt}}$ exponentially scales the complexity $\mathcal{N}(\mathcal{H}, \epsilon, |\cdot|)$. Meanwhile, the qubits count N and the operator norm $\|O\|$ polynomially scale the complexity of $\mathcal{N}(\mathcal{H}, \epsilon, |\cdot|)$. These observations suggest a succinct and direct way to compare the expressivity of QNNs with different quantum circuits. Moreover, the dependence of the expressivity of QNNs on the type of quantum gates (denoted by the term k) demonstrated that the expressivity of QNNs depends on the structure information of ansatz such as the location of different quantum gates and the types of the employed quantum gates. The expressivity measured by the covering number could provide practical guidance for designing the circuit structure of QNNs.

4.4.1.3 Generalization Error of QNNs

As the relation between generalization error and covering number is well established in statistical learning theory, we can directly obtain the generalization error bound with the above bounds of covering number following the same conventions.

Theorem 4.6 *Assume that the loss function ℓ defined in Eq. (4.38) is L_1-Lipschitz and upper bounded by a constant C, the QNN-based learning algorithm outputs a hypothesis $h_{\hat{\theta}}$ from the training dataset S of size n. Following the notations of risk $R_{\text{Gene}}(h_{\hat{\theta}}) = R(h_{\hat{\theta}}) - R_{\text{ERM}}(h_{\hat{\theta}})$ defined in Eq. (4.41), for $0 < \epsilon < 1/10$, with probability at least $1 - \delta$ with $\delta \in (0, 1)$, we have*

$$R_{\text{Gene}}(h_{\hat{\theta}}) \leq O\left(\frac{8L + c + 24L\sqrt{N_{gt}} \cdot 2^k}{\sqrt{n}}\right). \tag{4.57}$$

Proof Sketch of Theorem 4.6 Recall that the bound of generalization error in terms of Rademacher complexity has been established by Kakade et al. [48] as follows:

$$R_{\text{Gene}}(h_{\hat{\theta}}) \leq 2L_1 \mathcal{R}(\mathcal{H}_{QNN}) + 3C\sqrt{\frac{\ln(2/\delta)}{2n}}, \tag{4.58}$$

where $\mathcal{R}(\mathcal{H}_{QNN})$ represents the empirical Rademacher complexity of the hypothesis space of QNNs. Furthermore, the relationship between Rademacher complexity and covering number can be derived using the Dudley entropy integral bound [49], which is given by

$$\mathcal{R}(\mathcal{H}) \leq \inf_{\alpha > 0}\left(4\alpha + \frac{12}{\sqrt{n}}\int_{\alpha}^{1}\sqrt{\ln \mathcal{N}(\mathcal{H}_{|S}, \epsilon, \|\cdot\|_2)}d\epsilon\right), \tag{4.59}$$

where $\mathcal{H}_{|S}$ denotes the set of vectors formed by the hypothesis with n examples in the dataset S. In this regard, the generalization error bound in Eq. (4.57) could be obtained by combining the Eqs. (4.58) and (4.59) with direct but tedious calculations, which is omitted here. For details of the calculations, please refer to the proof of Theorem 2 in [46]. $\qquad\square$

The assumption used in this analysis is quite mild, as the loss functions in QNNs are generally Lipschitz continuous and can be bounded above by a constant C. This property has been broadly employed to understand the capability of QNNs. The results obtained have three key implications. First, the generalization bound exhibits an exponential dependence on the term k and a sublinear dependence on the number of trainable quantum gates N_{gt}. This observation reflects the quantum version of Occam's razor [50], where the parsimony of the output hypothesis implies greater predictive power. Second, increasing the number of training examples n improves the generalization bound. This suggests that incorporating more training data is essential for optimizing complex quantum circuits. Lastly, the sublinear dependence on N_{gt} may limit the ability to accurately assess the generalization performance of overparameterized QNNs [51]. Together, these implications provide valuable insights for designing more powerful QNNs.

4.4.2   Trainability of Quantum Neural Networks

The parameters in QNNs are often trained using gradient-based optimizers. As such, the magnitude of the gradient plays a crucial role in the trainability of QNNs. Specifically, large gradients are desirable, as they allow the loss function to decrease rapidly and consistently. However, this favorable property does not hold across a wide range of problem settings. In contrast, training QNNs usually encounters the barren plateau (BP) problem [52], i.e., *the variance of the gradient, on average, decreases exponentially as the number of qubits increases.* In this section, we first introduce an example demonstrating how quantum circuits that form unitary 2-designs [53] lead to BP and then discuss several techniques to avoid or mitigate this issue.

We begin by introducing some basic notations. For convenience, let L denote the number of parameters in the QNN $V(\boldsymbol{\theta})$. Consider the loss function defined as the measurement outcome of an N-qubit quantum state ρ after applying the QNN operation, i.e.,

$$f(\boldsymbol{\theta}) = \mathrm{Tr}\left[O V(\boldsymbol{\theta})\rho V(\boldsymbol{\theta})^{\dagger} \right]. \tag{4.60}$$

Then, the mathematical formulation of the BP phenomenon is given by

$$\mathbb{E}_{\mathcal{P}}\left[\frac{\partial f}{\partial \theta_k}\right] = 0, \quad \mathrm{Var}_{\mathcal{P}}\left[\frac{\partial f}{\partial \theta_k}\right] = \exp(-\alpha N) \cdot \beta, \tag{4.61}$$

where $\mathcal{P}$ represents the probability distribution of the quantum circuit and $\alpha, \beta > 0$ are constants. In the case where the circuit $V(\boldsymbol{\theta})$ has a random structure with a polynomial number of single-qubit rotations and CNOT or CZ gates in N, a uniform distribution over the parameter space can approximate a 2-design for the unitary $V(\boldsymbol{\theta})$ [54, 55]. Moreover, a unitary sampled from an exact 2-design exhibits the following statistical properties.

Fact 4.3 ([56]) *Let $\{W_y\}_{y\in Y} \subset \mathcal{U}(d)$ form a unitary 2-design, and let $A, B, C, D : \mathcal{H}_w \to \mathcal{H}_w$ be arbitrary linear operator. Then*

$$\frac{1}{|Y|} \sum_{y\in Y} Tr[W_y A W_y^{\dagger} B] = \frac{Tr[A]\, Tr[B]}{d}, \tag{4.62}$$

$$\frac{1}{|Y|} \sum_{y \in Y} Tr[W_y A W_y^\dagger B]\, Tr[W_y C W_y^\dagger D]$$

$$= \frac{1}{d^2 - 1}\left(Tr[AC]\, Tr[BD] + Tr[A]\, Tr[B]\, Tr[C]\, Tr[D]\right)$$

$$- \frac{1}{d(d^2 - 1)}\left(Tr[A]\, Tr[C]\, Tr[BD] + Tr[AC]\, Tr[B]\, Tr[D]\right), \tag{4.63}$$

$$\frac{1}{|Y|} \sum_{y \in Y} Tr[W_y A W_y^\dagger B W_y C W_y^\dagger D]$$

$$= \frac{1}{d^2 - 1}\left(Tr[A]\, Tr[C]\, Tr[BD] + Tr[AC]\, Tr[B]\, Tr[D]\right)$$

$$- \frac{1}{d(d^2 - 1)}\left(Tr[AC]\, Tr[BD] + Tr[A]\, Tr[B]\, Tr[C]\, Tr[D]\right). \tag{4.64}$$

Fact 4.3 can be derived from Facts B.1 and B.2 in Appendix B, which provides a more detailed discussion of unitary designs, potentially of independent interest. By applying Fact 4.3, it can be shown that QNNs with quantum circuits forming 2-designs exhibit barren plateau loss landscapes.

Theorem 4.7 (Adapted from [52]) *Consider the loss function given in Eq. (4.60), where the QNN $V(\boldsymbol{\theta}) = \prod_{j=1}^{L} V_j(\boldsymbol{\theta}_j) W_j$ with fixed gate W_j and variational gate $V_j(\boldsymbol{\theta}_j) = \exp(-i\boldsymbol{\theta}_j H_j/2)$. Suppose all hermitian matrices $\{H_j\}$ are traceless. For an integer $k \in [1, L]$, denote $U_- = \prod_{j=1}^{k-1} V_j(\boldsymbol{\theta}_j) W_j$ and $U_+ = \prod_{j=k+1}^{L} V_j(\boldsymbol{\theta}_j) W_j$. Then, if both U_- and U_+ form 2-designs, there is*

$$\mathbb{E}\left[\frac{\partial f}{\partial \boldsymbol{\theta}_k}\right] = 0, \quad \mathrm{Var}\left[\frac{\partial f}{\partial \boldsymbol{\theta}_k}\right] \approx \frac{1}{2^{3N+1}}\, Tr\left[O^2\right] Tr\left[\rho^2\right] Tr\left[H_j^2\right]. \tag{4.65}$$

Proof (Proof of Theorem 4.7)

By using notations U_- and U_+, the function $f(\boldsymbol{\theta})$ in Eq. (4.60) can be expressed as

$$f = Tr\left[O V \rho V^\dagger\right]$$

$$= Tr\left[O U_- V_k W_k U_+ \rho U_+^\dagger W_k^\dagger V_k^\dagger U_-^\dagger\right]$$

$$= Tr\left[U_-^\dagger O U_- V_k W_k U_+ \rho U_+^\dagger W_k^\dagger V_k^\dagger\right]$$

$$= Tr\left[O' \exp(-i\boldsymbol{\theta}_k H_k/2) \rho' \exp(i\boldsymbol{\theta}_k H_k/2)\right],$$

where $O' := U_-^\dagger O U_-$ and $\rho' := W_k U_+ \rho U_+^\dagger W_k^\dagger$. Thus, the gradient could be calculated as

$$\frac{\partial f}{\partial \theta_k} = \frac{i}{2} \operatorname{Tr}\left[O'\left[V_k \rho' V_k^\dagger, H_k\right]\right].$$

The expectation of the gradient is zero since

$$\underset{U_+, U_-}{\mathbb{E}} \frac{\partial f}{\partial \theta_k} = \mathbb{E}\frac{i}{2} \operatorname{Tr}\left[O'\left[V_k \rho' V_k^\dagger, H_k\right]\right]$$

$$= \underset{U_+, U_-}{\mathbb{E}} \frac{i}{2} \operatorname{Tr}\left[U_-^\dagger O U_-\left[V_k \rho' V_k^\dagger, H_k\right]\right]$$

$$= \underset{U_+}{\mathbb{E}} \frac{i}{2^{N+1}} \operatorname{Tr}\left[O\right] \operatorname{Tr}\left[\left[V_k \rho' V_k^\dagger, H_k\right]\right] \tag{4.66}$$

$$= 0, \tag{4.67}$$

where Eq. (4.66) follows from Eqs. (4.62) and (4.67) is derived by noticing $\operatorname{Tr}[[A, B]] = \operatorname{Tr}[AB - BA] = 0$. Therefore, the variance of the gradient equals to the expectation of its square, i.e.,

$$\underset{U_+, U_-}{\operatorname{Var}}\left[\frac{\partial f}{\partial \theta_k}\right] = \underset{U_+, U_-}{\mathbb{E}}\left[\frac{\partial f}{\partial \theta_k}\right]^2$$

$$= -\frac{1}{4}\mathbb{E}_{U_+, U_-} \operatorname{Tr}\left[U_-^\dagger O U_-\left[V_k \rho' V_k^\dagger, H_k\right]\right]^2$$

$$= -\frac{1}{4 \times (2^{2N} - 1)} \underset{U_+}{\mathbb{E}} \operatorname{Tr}\left[O^2\right] \operatorname{Tr}\left[\left[V_k \rho' V_k^\dagger, H_k\right]^2\right]$$

$$+ \frac{1}{2^{N+2}\left(2^{2N} - 1\right)} \underset{U_+}{\mathbb{E}} \operatorname{Tr}\left[O\right]^2 \operatorname{Tr}\left[\left[V_k \rho' V_k^\dagger, H_k\right]^2\right] \tag{4.68}$$

$$= -\frac{1}{4 \times \left(2^{2N} - 1\right)} \operatorname{Tr}\left[O^2\right] \underset{U_+}{\mathbb{E}} \operatorname{Tr}\left[\left[V_k \rho' V_k^\dagger, H_k\right]^2\right], \tag{4.69}$$

where Eq. (4.68) follows from Eqs. (4.63) and (4.69) follows from $\operatorname{Tr}[O] = 0$. Further, it can be shown that

$$\underset{U_+}{\mathbb{E}} \operatorname{Tr}\left[\left[V_k \rho' V_k^\dagger, H_k\right]^2\right]$$

$$= 2 \underset{U_+}{\mathbb{E}} \operatorname{Tr}\left[\left(V_k \rho' V_k^\dagger H_k\right)^2\right] - 2 \underset{U_+}{\mathbb{E}} \operatorname{Tr}\left[\left(V_k \rho' V_k^\dagger\right)^2 (H_k)^2\right]$$

$$= 2 \underset{U_+}{\mathbb{E}} \operatorname{Tr}\left[\left(V_k W_k U_+ \rho U_+^\dagger W_k^\dagger V_k^\dagger H_k\right)^2\right]$$

$$-2 \underset{U_+}{\mathbb{E}} \mathrm{Tr}\left[\left(V_k W_k U_+ \rho U_+^\dagger W_k^\dagger V_k^\dagger\right)^2 (H_k)^2\right]$$

$$= \frac{2}{2^{2N}-1}\left\{\mathrm{Tr}[\rho]^2\,\mathrm{Tr}[H_k^2] + \mathrm{Tr}\left[\rho^2\right]\mathrm{Tr}[H_k]^2\right\}$$

$$-\frac{2}{2^N(2^{2N}-1)}\left\{\mathrm{Tr}\left[\rho^2\right]\mathrm{Tr}\left[H_k^2\right] + \mathrm{Tr}[\rho]^2\,\mathrm{Tr}[H_k]^2\right\}$$

$$-\frac{2}{2^N}\mathrm{Tr}\left[H_k^2\right]\mathrm{Tr}\left[\rho^2\right] \tag{4.70}$$

$$\approx -\frac{2^{N+1}}{2^{2N}-1}\mathrm{Tr}\left[H_k^2\right]\mathrm{Tr}\left[\rho^2\right], \tag{4.71}$$

where Eq. (4.70) follows from Eqs. (4.62) and (4.64). Equation (4.71) is derived by ignoring minor terms and using $\mathrm{Tr}[H_k] = 0$. Combining Eqs. (4.69) and (4.71), it can be shown that

$$\underset{U_+,U_-}{\mathrm{Var}}\left[\frac{\partial f}{\partial \boldsymbol{\theta}_k}\right] \approx \frac{2^N}{2\times\left(2^{2N}-1\right)^2}\mathrm{Tr}\left[O^2\right]\mathrm{Tr}\left[H_k^2\right]\mathrm{Tr}\left[\rho^2\right]$$

$$\approx \frac{1}{2^{3N+1}}\mathrm{Tr}\left[O^2\right]\mathrm{Tr}\left[\rho^2\right]\mathrm{Tr}\left[H_k^2\right].$$

Thus, Theorem 4.7 is proved.

$\square$

Remark
The influence of barren plateau can be categorized into three folds.

1. **Training efficiency.** Exponentially small gradients imply that the training of QNNs with gradient-based optimizers may require exponential numbers of iterations to converge.
2. **Optimization effectiveness.** The calculation of the gradient via the parameter-shift rule would introduce unbiased statistical noise due to finite shot numbers. A small gradient could be vulnerable to these measurement noises, which may induce additional barriers in optimizations.
3. **Quantum advantage.** QNNs are expected to exhibit quantum advantages by employing intermediate to large numbers of qubits. However, the barren plateau phenomenon may induce exponential training steps, which can offset potential quantum advantages.

Since the barren plateau could seriously affect the trainability of scaled QNNs and raise concerns about the utility of QNNs for achieving quantum advantages, researchers have been focused on developing techniques to address this problem. Existing efforts include specific architecture design [56–58], parameter initialization schemes [59, 60], and advanced training protocols [61, 62]. Here, we briefly introduce two related results with theoretical guarantees.

Fact 4.4 (Shallow Hardware-Efficient Circuits are BP-Free, Informal Version Adapted from [56]) *Suppose the observable has a local form in the Pauli basis decomposition. For QNNs employing N-qubit shallow hardware-efficient circuits with logarithmic depths, the variance of the gradient has the lower bound:*

$$\mathrm{Var}\left[\frac{\partial f}{\partial \boldsymbol{\theta}_k}\right] \geq \Omega\left(\frac{1}{\mathrm{poly}(N)}\right).$$
(4.72)

Fact 4.5 (Gaussian Initializations Help to Escape the BP Region, Informal Version Adapted from [60]) *Suppose the observable is the tensor product of Pauli matrices $\sigma_i = \sigma_{i_1} \otimes \cdots \otimes \sigma_{i_N}$, where the number of nonidentity matrices in $\{\sigma_{i_1}, \cdots, \sigma_{i_N}\}$ is S. For QNNs employing N-qubit shallow hardware-efficient circuits with the depth L, the gradient norm has the lower bound:*

$$\mathbb{E}_{\boldsymbol{\theta}} \|\nabla_{\boldsymbol{\theta}} f\|^2 \geq \frac{L}{S^S(L+2)^{S+1}} \mathrm{Tr}\left[\sigma_j \rho_{\mathrm{in}}\right]^2,$$
(4.73)

where S is the number of nonzero elements in $\boldsymbol{i}$, and the index $\boldsymbol{j} = (j_1, j_2, \cdots, j_N)$ such that $j_m = 0, \forall i_m = 0$ and $j_m = 3, \forall i_m \neq 0$. The expectation is taken with the Gaussian distribution $\mathcal{N}\left(0, \frac{1}{4S(L+2)}\right)$ for the parameters $\boldsymbol{\theta}$.

4.5 Code Demonstration

This section provides hands-on demonstrations of QNNs for both discriminative and generative tasks, which illustrate practical implementations of quantum classifiers and quantum patch GAN. Each subsection corresponds to a specific application, offering a step-by-step explanation and code walkthrough.

4.5.1 Quantum Classifier

We now demonstrate how to utilize QNN to solve discriminative tasks, specifically a binary classification problem based on the Wine dataset, which is widely used in machine learning benchmarks and classification tasks. The major steps are outlined below:

Step 1 Load and preprocess the dataset.

Step 2 Implement a quantum read-in protocol to encode classical data into quantum states.

Step 3 Construct a parameterized quantum circuit model to process the input quantum states.

Step 4 Train and test the QNN to evaluate its performance.

We first import all the necessary libraries:

```python
import sklearn
import sklearn.datasets
import pennylane as qml
from pennylane import numpy as np
from pennylane.optimize import AdamOptimizer
import matplotlib.pyplot as plt
```

Step 1: Dataset Preparation The Wine dataset is prepared for the classification task. For simplicity, the focus is on the first two classes of the Wine dataset. The dataset consists of 13 attributes per sample, each with a distinct range. The normalization is applied to rescale these attributes to the interval $[0, \pi]$. Furthermore, the labels are remapped from $\{0, 1\}$ to $\{-1, 1\}$ to align with the output range of the quantum circuit model. The dataset is split into training and test sets to fairly evaluate the classifier.

```python
def load_wine(split_ratio = 0.5):
    feat, label = sklearn.datasets.load_wine(return_X_y=True)

    # normalization
    feat = np.pi * (feat - np.min(feat, axis=0, keepdims=True)
        ) / np.ptp(feat, axis=0, keepdims=True)

    index_c0 = label == 0
    index_c1 = label == 1

    label = label * 2 - 1

    n_c0 = sum(index_c0)
    n_c1 = sum(index_c1)

    X_train = np.concatenate((feat[index_c0][:int(split_ratio*
        n_c0)], feat[index_c1][:int(split_ratio*n_c1)]), axis
        =0)
    y_train = np.concatenate((label[index_c0][:int(split_ratio
        *n_c0)], label[index_c1][:int(split_ratio*n_c1)]),
        axis=0)
    X_test = np.concatenate((feat[index_c0][int(split_ratio*
        n_c0):], feat[index_c1][int(split_ratio*n_c1):]), axis
        =0)
    y_test = np.concatenate((label[index_c0][int(split_ratio*
        n_c0):], label[index_c1][int(split_ratio*n_c1):]),
        axis=0)
```

```
     return X_train, y_train, X_test, y_test
X_train, y_train, X_test, y_test = load_wine()
```

To better understand the dataset, t-SNE is employed to visualize its distribution. As shown in Fig. 4.12, each data point is projected into a 2D space for visualization, with distinct colors representing different classes.

```python
def visualize_dataset(X, y):
    from sklearn.manifold import TSNE

    tsne = TSNE(n_components=2, random_state=42, perplexity
        =30)

    wine_tsne = tsne.fit_transform(X)
    for label in np.unique(y):
        indices = y == label
        plt.scatter(wine_tsne[indices, 0], wine_tsne[indices,
            1], edgecolor='black', cmap='coolwarm', s=20,
            label=f'Class {label}')

    # Add labels and legend
    plt.title("t-SNE Visualization of Wine dataset (two
        classes)")
    plt.xlabel("t-SNE Dimension 1")
    plt.ylabel("t-SNE Dimension 2")
    plt.legend()

    plt.tight_layout()
    plt.show()
visualize_dataset(X_train, y_train)
```

Step 2: Data Encoding To encode the 13 attributes of the Wine dataset into a quantum system, angle encoding introduced in Sect. 2.3.1.3 is used, followed by a layer of CNOT gates acting on neighboring qubits to introduce entanglement.

```python
def data_encoding(x):
    n_qubit = len(x)
    qml.AngleEmbedding(features =x , wires = range(n_qubit) ,
        rotation ="X")
    for i in range(n_qubit):
        if i+1 < n_qubit:
            qml.CNOT(wires=[i, i+1])
```

Step 3: Building Quantum Classifier With the data encoding in place, a quantum binary classifier is constructed. The circuit model is composed of multiple layers, where each layer includes parameterized single-qubit rotation gates with trainable angles, followed by a block of nonparametric CNOT gates to introduce entanglement among qubits. To read-out the category information of each input sample from the prepared quantum state, the expectation value of the Pauli-Z operator on the first qubit is calculated.

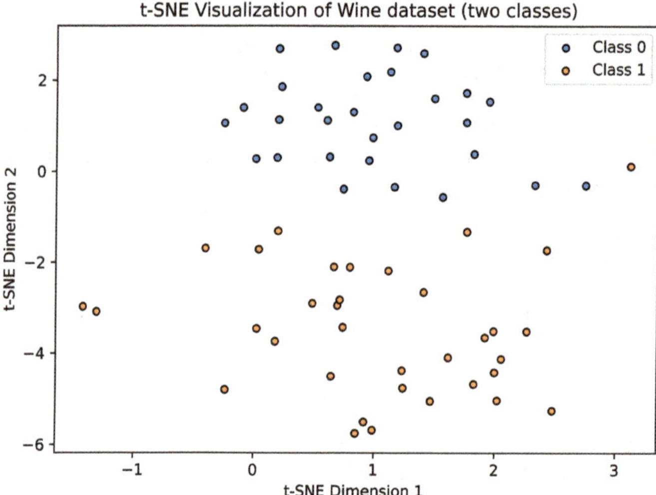

Fig. 4.12 T-SNE visualization of Wine dataset of the first two classes

```
def classifier(param, x=None):
    data_encoding(x)
    n_layer, n_qubit = param.shape[0], param.shape[1]
    for i in range(n_layer):
        for j in range(n_qubit):
            qml.Rot(param[i, j, 0], param[i, j, 1], param[i, j
                , 2], wires=j)
        for j in range(n_qubit):
            if j+1 < n_qubit:
                qml.CNOT(wires=[j, j+1])
    return qml.expval(qml.PauliZ(0))

n_qubit = X_train.shape[1]
dev = qml.device('default.qubit', wires=n_qubit)
circuit = qml.QNode(classifier, dev)
```

The whole quantum circuit of two layers is visualized by drawing the diagram, as shown in Fig. 4.13.

```
fig, ax = qml.draw_mpl(circuit)(np.pi * np.random.randn(2,
    n_qubit, 3), X_train[0])
fig.show()
```

Step 4: Training and Evaluation of Quantum Classifier With the data and circuit model ready, we now move to the optimization of the quantum classifier. The mean squared error (MSE) is used as the loss function. The goal is to minimize the difference between the predicted and actual labels over the training set.

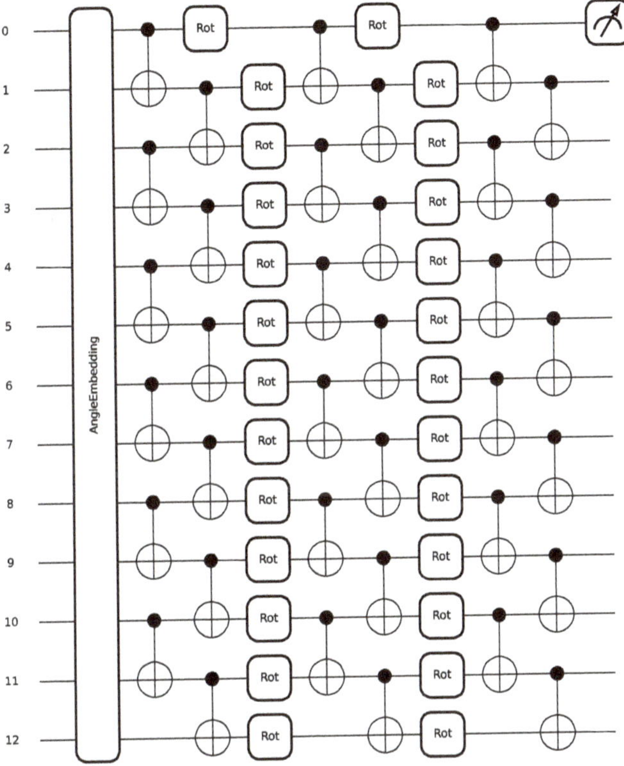

Fig. 4.13 The circuit diagram of the quantum classifier

```python
def mse_loss(predict, label):
    return np.mean((predict - label)**2)

def cost(param, circuit, X, y):
    exp = []
    for i in range(len(X)):
        pred = circuit(param, x=X[i])
        exp.append(pred)
    return mse_loss(np.array(exp), y)
```

To evaluate the performance of the quantum classifier, classification accuracy is exploited as the metric. Specifically, if the sign of the read-out result matches the corresponding label, the prediction is considered correct; otherwise, it is deemed incorrect. The accuracy is then calculated as the proportion of correctly classified samples out of the total.

```python
def accuracy(predicts, labels):
    assert len(predicts) == len(labels)
    return np.sum((np.sign(predicts)*labels+1)/2)/len(predicts
        )
```

The Adam optimizer is utilized to minimize the loss function. To ensure efficient computation, the dataset is divided into smaller batches for each training iteration. At the end of each epoch, both the training and test losses, along with the classification accuracy, are recorded to track the model's performance.

```python
lr = 0.01
opt = AdamOptimizer(lr)
batch_size = 4

n_epoch = 50
cost_train, cost_test, acc_train, acc_test = [], [], [], []
for i in range(n_epoch):
    index = np.random.permutation(len(X_train))
    feat_train, label_train = X_train[index], y_train[index]
    for j in range(0, len(X_train), batch_size):
        feat_train_batch = feat_train[j*batch_size:(j+1)*
            batch_size]
        label_train_batch = label_train[j*batch_size:(j+1)*
            batch_size]
        param = opt.step(lambda v: cost(v, circuit,
            feat_train_batch, label_train_batch), param)

    # compute cost
    cost_train.append(cost(param, circuit, X_train, y_train))
    cost_test.append(cost(param, circuit, X_test, y_test))

    # compute accuracy
    pred_train = []
    for j in range(len(X_train)):
        pred_train.append(circuit(param, x=X_train[j]))
    acc_train.append(accuracy(np.array(pred_train), y_train))

    pred_test = []
    for j in range(len(X_test)):
        pred_test.append(circuit(param, x=X_test[j]))
    acc_test.append(accuracy(np.array(pred_test), y_test))
```

After training the QNN, the training and test cost, as well as the accuracy over epochs, can be visualized using the following code. As demonstrated in Fig. 4.14, this QNN achieves a test accuracy exceeding 80%.

```python
plt.figure(figsize=(12, 6))

plt.subplot(1, 2, 1)
epochs = np.arange(n_epoch) + 1
plt.plot(epochs, cost_train, label='Training_Cost', marker='o'
    )
plt.plot(epochs, cost_test, label='Test_Cost', marker='o')
plt.title('Cost_Over_Epochs')
plt.xlabel('Epochs')
plt.ylabel('Cost')
plt.legend()
plt.grid()
```

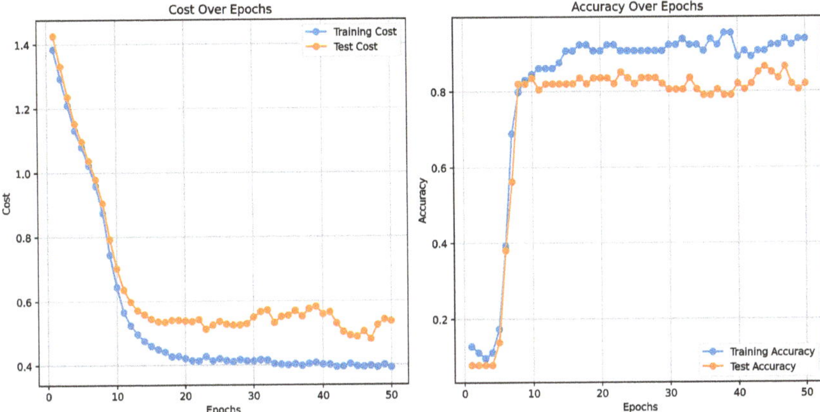

Fig. 4.14   The training curve of the quantum classifier

```python
# Plot training and test accuracy
plt.subplot(1, 2, 2)
plt.plot(epochs, acc_train, label='Training Accuracy', marker=
    'o')
plt.plot(epochs, acc_test, label='Test Accuracy', marker='o')
plt.title('Accuracy Over Epochs')
plt.xlabel('Epochs')
plt.ylabel('Accuracy')
plt.legend()
plt.grid()

plt.tight_layout()
plt.show()
```

To better understand the performance of the quantum classifier in comparison to a classical counterpart, an MLP for the same classification task is also implemented.

```python
class MLP(nn.Module):
    def __init__(self):
        super(MLP, self).__init__()
        self.fc1 = nn.Linear(13, 10)   # 13 input features to
            10 neurons
        self.fc2 = nn.Linear(10, 2)    # 10 neurons to 2 output
            classes

    def forward(self, x):
        x = torch.relu(self.fc1(x))
        x = self.fc2(x)
        return x
```

In this implementation, the number of hidden neurons in the MLP is set to 10, resulting in approximately 150 trainable parameters, which is comparable to the

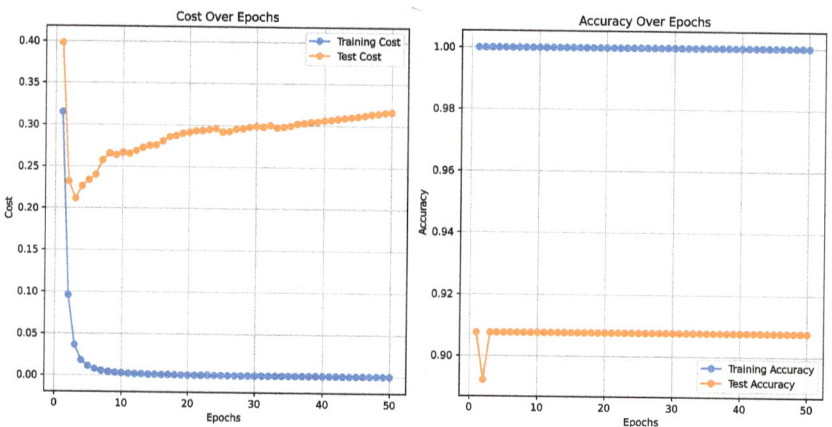

Fig. 4.15 The training curve of the classical MLP

number of parameters in the QNN. All other hyperparameters, such as the optimizer and learning rate, remain the same to ensure a fair comparison. The training curve for the MLP is shown in Fig. 4.15. The MLP converges quickly, achieving 100% training accuracy and over 90% test accuracy.

While the MLP currently demonstrates superior accuracy, the QNN still provides a promising alternative. It is important to note that the performance of the QNN could potentially be further enhanced by employing more advanced read-in protocols, as discussed in Sect. 2.3.1, which could enable more efficient and expressive representations of the input data. Additionally, optimizing the circuit design, such as adjusting the arrangement of layers or introducing more complex parameterized gates to increase the model's capacity, as highlighted in Sect. 4.6, could further improve the QNN's ability to capture intricate patterns in the dataset.

4.5.2 Quantum Patch GAN

We next demonstrate how to implement a quantum patch GAN introduced in Sect. 4.3.3 for the generation of handwritten digits of five. To evaluate the performance of the generative model, the *Fréchet Distance (FD)*, a widely used metric for quantifying the difference between two distributions, is employed. The metric is formally defined as

$$FD = \left| \mu_r - \mu_g \right|^2 + \mathrm{Tr}(\Sigma_r + \Sigma_g - 2(\Sigma_r \Sigma_g)^{\frac{1}{2}}), \tag{4.74}$$

where μ_r and μ_g are the mean feature vectors of real and generated images and Σ_r and Σ_g are the covariance matrices of the feature vectors of real and generated images. A lower FD value indicates that the generated images closely resemble the real ones, while a higher FD suggests a greater discrepancy between the distributions. This metric provides an effective way to quantitatively evaluate the quality of the generated samples. In addition to FD, other commonly used metrics for evaluating the quality of generated images include inception score and kernel inception distance.

The whole pipeline includes the following steps:

Step 1 Load and preprocess the dataset.
Step 2 Build the classical discriminator.
Step 3 Build the quantum generator.
Step 4 Train the quantum patch GAN.
Step 5 Visualize the generated images.

We begin by importing the required libraries:

```python
import torch
import torch.nn as nn
import torch.optim as optim
from torch.utils.data import Dataset, DataLoader

import numpy as np
import matplotlib.pyplot as plt
import pennylane as qml
import math
```

Step 1: Dataset Preparation We will use the Optical Recognition of Handwritten Digits dataset (`optdigits`), where each data point represents an 8×8 grayscale image. The following code defines a custom dataset class to load and process the data.

```python
class OptdigitsData(Dataset):
    def __init__(self, data_path, label):
        """
        Dataset class for Optical Recognition of Handwritten
            Digits.
        """
        super().__init__()

        self.data = []
        with open(data_path, 'r') as f:
            for line in f.readlines():
                if int(line.strip().split(',')[-1]) == label:
                    # Normalize image pixel values from [0,16)
                        to [0, 1)
                    image = [int(pixel)/16 for pixel in line.
                        strip().split(',')[:-1]]
                    image = np.array(image, dtype=np.float32).
                        reshape(8, 8)
                    self.data.append(image)
        self.label = label
```

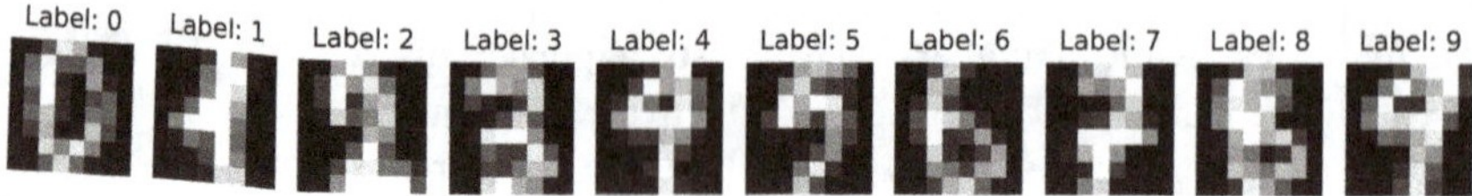

Fig. 4.16 Samples in the dataset optdigits

```python
def __len__(self):
    return len(self.data)

def __getitem__(self, index):
    return torch.from_numpy(self.data[index]), self.label
```

After defining the dataset, we can visualize a few examples as shown in Fig. 4.16 to better understand the structure of the dataset.

```python
def visualize_dataset(data_path):
    """
    Visualizes the dataset by displaying examples for each
        digit label.
    """
    plt.figure(figsize=(10, 5))
    for i in range(10):
        plt.subplot(1, 10, i + 1)
        data = OptdigitsData(data_path, label=i)
        plt.imshow(data[0][0], cmap='gray')
        plt.title(f"Label:_{i}")
        plt.axis('off')
    plt.tight_layout()
    plt.show()

visualize_dataset('code/chapter5_qnn/data/optdigits.tra')
```

Step 2: Building the Classical Discriminator The discriminator is a classical neural network responsible for distinguishing real images from fake ones. It consists of fully connected layers with ReLU activations. The final output is passed through a Sigmoid function, which scales the output to the range $(0, 1)$, serving as a probabilistic indicator of whether the input image is real or fake.

```python
class ClassicalDiscriminator(nn.Module):
    """
    A classical discriminator for classifying real and fake
        images.
    """

    def __init__(self, input_shape):
        super().__init__()
        self.model = nn.Sequential(
            nn.Flatten(),
            nn.Linear(int(np.prod(input_shape)), 256),
            nn.ReLU(),
            nn.Dropout(),
```

```
12          nn.Linear(256, 128),
13          nn.ReLU(),
14          nn.Dropout(),
15          nn.Linear(128, 1),
16          nn.Sigmoid()
17      )
18
19  def forward(self, img):
20      return self.model(img)
```

Step 3: Defining the Quantum Patch Generator The generator in the quantum patch GAN consists of parameterized quantum circuits (PQC). These circuits are responsible for generating patches of the target image. Specifically, the PQC follows the layout in Fig. 4.7, which applies layers of single-qubit rotation gates and entangling gates to the latent state.

```
1  def PQC(params):
2      n_layer, n_qubit = params.shape[0], params.shape[1]
3      for i in range(n_layer):
4          for j in range(n_qubit):
5              qml.Rot(params[i, j, 0], params[i, j, 1], params[i
                  , j, 2], wires=j)
6          # Control Z gates
7          for j in range(n_qubit - 1):
8              qml.CZ(wires=[j, j + 1])
```

Then, we implement the quantum generator for each patch of an image, i.e., sub-generators. The sub-generator transforms the latent variable z into a latent quantum state $|z\rangle$, applies a PQC, performs partial measurements on the ancillary system $\mathcal{A}$, and outputs the probabilities of each computational basis state in the remaining system, which correspond to the generated pixel values.

```
1   def QuantumGenerator(params, z=None, n_qubit_a=1):
2       n_qubit = params.shape[1]
3
4       # angle encoding of latent state z
5       for i in range(n_qubit):
6           qml.RY(z[i], wires=i)
7
8       PQC(params)
9
10      # partial measurement on the ancillary qubits
11      qml.measure(wires=n_qubit-1)
12      return qml.probs(wires=range(n_qubit-n_qubit_a))
```

Using the sub-generators, the quantum patch generator combines the output patches from multiple sub-generators to construct the complete image.

```python
class PatchQuantumGenerator(nn.Module):
    """
    Combines patches generated by quantum circuits into full
        images.
    """
    def __init__(self, qnode_generator, n_generator, n_qubit,
        n_qubit_a, n_layer):
        super().__init__()

        self.params_generator = nn.ParameterList([
            nn.Parameter(torch.rand((n_layer, n_qubit, 3)),
                requires_grad=True) for _ in range(n_generator
                )
        ])
        self.qnode_generator = qnode_generator
        self.n_qubit_a = n_qubit_a

    def forward(self, zs):
        images = []
        for z in zs:
            patches = []
            for params in self.params_generator:
                patch = self.qnode_generator(params, z=z,
                    n_qubit_a=self.n_qubit_a).float()

                # post-processing: min-max scaling
                patch = (patch - patch.min()) / (patch.max() -
                    patch.min() + 1e-8)

                patches.append(patch.unsqueeze(0))
            patches = torch.cat(patches, dim=0)
            images.append(patches.flatten().unsqueeze(0))
        return torch.cat(images, dim=0)
```

Step 4: Training the Quantum Patch GAN With the dataset and models ready,
we initialize the quantum generator, classical discriminator, and their optimizers.

```python
# Hyperparameters
torch.manual_seed(0)
image_width = 8
image_height = 8
n_generator = 4
n_qubit_d = int(np.log2((image_width * image_height) //
    n_generator))
n_qubit_a = 1
n_qubit = n_qubit_d + n_qubit_a
n_layer = 6

# Quantum device
dev = qml.device("lightning.qubit", wires=n_qubit)
qnode_generator = qml.QNode(QuantumGenerator, dev)
```

```python
# Initialize generator and discriminator
discriminator = ClassicalDiscriminator([image_height,
    image_width])
discriminator.train()
generator = PatchQuantumGenerator(qnode_generator, n_generator
    , n_qubit, n_qubit_a, n_layer)
generator.train()

# Optimizers
lr_generator = 0.3
lr_discriminator = 1e-2
opt_discriminator = optim.SGD(discriminator.parameters(), lr=
    lr_discriminator)
opt_generator = optim.SGD(generator.parameters(), lr=
    lr_generator)

# Construct dataset and dataloader
batch_size = 4
dataset = OptdigitsData('code/chapter5_qnn/data/optdigits.tra'
    , label=5)
dataloader = DataLoader(dataset, batch_size=batch_size,
    shuffle=True, drop_last=True)

# Loss function
loss_fn = nn.BCELoss()
labels_real = torch.ones(batch_size, dtype=torch.float)
labels_fake = torch.zeros(batch_size, dtype=torch.float)

# Testing setup
n_test = 10
z_test = torch.rand(n_test, n_qubit) * math.pi
```

The GAN training process involves alternating updates for the discriminator and
the generator. The discriminator is trained to distinguish between real and fake
images, while the generator learns to create images that can successfully deceive
the discriminator. To track the generator's progress during training, the generated
images are saved at the end of each epoch.

```python
n_epoch = 10
record = {}
for i in range(n_epoch):
    for data, _ in dataloader:

        zs = torch.rand(batch_size, n_qubit) * math.pi
        image_fake = generator(zs)

        # Training the discriminator
        discriminator.zero_grad()
        pred_fake = discriminator(image_fake.detach())
        pred_real = discriminator(data)

```

```
14      loss_discriminator = loss_fn(pred_fake.squeeze(),
            labels_fake) + loss_fn(pred_real.squeeze(),
            labels_real)
15      loss_discriminator.backward()
16      opt_discriminator.step()
17
18      # Training the generator
19      generator.zero_grad()
20      pred_fake = discriminator(image_fake)
21      loss_generator = loss_fn(pred_fake.squeeze(),
            labels_real)
22      loss_generator.backward()
23      opt_generator.step()
24
25    print(f'The {i}-th epoch: discriminator loss={
          loss_discriminator: 0.3f}, generator loss={
          loss_generator: 0.3f}')
26
27    # test
28    generator.eval()
29    image_generated = generator(z_test).view(n_test,
          image_height, image_width).detach()
30
31    record[str(i)] = {
32        'loss_discriminator': loss_discriminator.item(),
33        'loss_generator': loss_generator.item(),
34        'image_generated': image_generated.numpy().tolist()
35    }
36    generator.train()
```

Step 5: Visualizing the Generated Images After training, the images generated by the quantum generator are visualized in Fig. 4.17 to evaluate its performance. These visualizations allow us to see how well the model has learned to produce realistic image patches.

```
1  n_epochs_to_visualize = len(record) // 2
2  n_images_per_epoch = 10
3
4  fig, axes = plt.subplots(n_epochs_to_visualize,
       n_images_per_epoch, figsize=(n_images_per_epoch,
       n_epochs_to_visualize))
5
6  # Iterate through the recorded epochs and visualize generated
       images
7  for epoch_idx, (epoch, data) in enumerate(record.items()):
8      if epoch_idx % 2 == 1:
9          continue
10     images = np.array(data['image_generated'])
11
12     for img_idx in range(n_images_per_epoch):
13         ax = axes[epoch_idx // 2, img_idx]
14         ax.imshow(images[img_idx], cmap='gray')
```

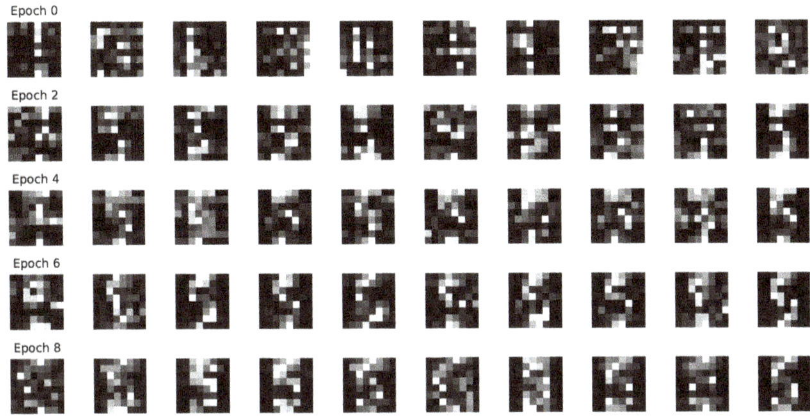

Fig. 4.17 The images generated by quantum patch GAN during training

```python
    ax.axis('off')

    # Add epoch information to the title of each row
    if img_idx == 0:
        ax.set_title(f"Epoch_{epoch}", fontsize=10)

plt.tight_layout()
plt.show()
```

To benchmark the performance of the quantum patch GAN, a classical patch GAN is implemented, where both the generator and discriminator are entirely classical. Specifically, each patch is generated using an MLP with independent parameters, and the final image is obtained by assembling all generated patches. This ensures a fair comparison between quantum and classical generative models.

```python
class ClassicalPatchGenerator(nn.Module):
    def __init__(self, latent_dim, patch_size, n_patches):
        super().__init__()
        self.latent_dim = latent_dim
        self.patch_size = patch_size
        self.n_patches = n_patches

        # MLP to generate each patch
        self.patch_generators = nn.ModuleList([nn.Sequential(
            nn.Linear(latent_dim, 5),  # Input layer
            nn.ReLU(),
            nn.Linear(5, patch_size * patch_size),  # Output
                layer (patch)
            nn.Sigmoid()  # Output in range [0, 1]
        ) for _ in range(n_patches)])

    def forward(self, zs):
        images = []
```

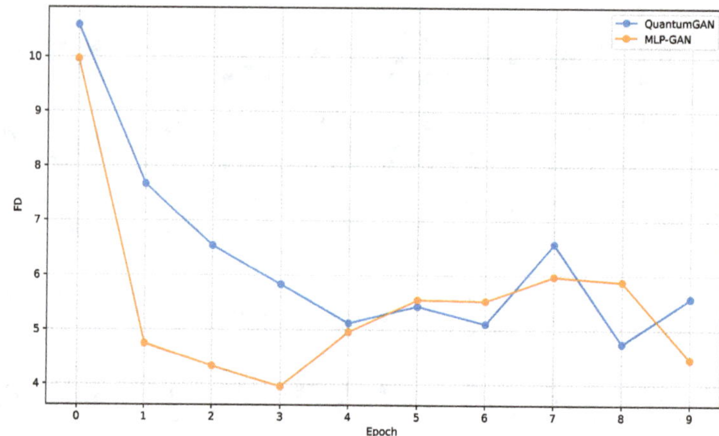

Fig. 4.18 FD values achieved by quantum patch GAN ("QuantumGAN") and classical patch GAN ("MLP-GAN")

```
18      for z in zs:
19          patches = []
20          for i in range(self.n_patches):
21              patch = self.patch_generators[i](z)   #
                    Generate a patch
22              patch = patch.view(self.patch_size, self.
                    patch_size)   # Reshape to patch size
23              patches.append(patch.unsqueeze(0))   # Add
                    batch dimension
24          patches = torch.cat(patches, dim=0)   # Combine
                patches
25          images.append(patches.flatten().unsqueeze(0))   #
                Flatten and add batch dimension
26      return torch.cat(images, dim=0)   # Combine all images
```

This choice of MLP architecture ensures that the number of trainable parameters in the classical GAN is comparable to that of the quantum GAN, making the comparison meaningful. With all other experimental settings unchanged, the same generation task is conducted using this MLP-based GAN (MLP-GAN). The FD values achieved by both models are presented in Fig. 4.18. The results show that the quantum GAN exhibits competitive performance compared to its classical counterpart, demonstrating the potential of quantum-enhanced generative models.

4.6   Bibliographic Remarks

Quantum neural networks (QNNs) are a major focus in quantum machine learning. They show promise for both classifying data (discriminative tasks) and creating new data (generative tasks). For classification, QNNs use quantum systems that can capture complicated patterns in data by working in high-dimensional spaces. For generating data, QNNs use flexible quantum circuits to create patterns that may be too difficult for classical models to handle. Although these approaches use similar learning tricks, each brings its own set of challenges in how the models are built, tested, and used in practice. In the remainder of this section, we briefly review recent advances in QNNs. Interested readers can refer to Ablayev et al. [63], Li and Deng [64], Massoli et al. [65], Li et al. [66], Tian et al. [67], and Wang and Liu [34] for a more comprehensive review.

4.6.1 Discriminative Learning with QNN

QNNs for classifying data have quickly become one of the busiest research areas in quantum machine learning. They may help find better ways to represent features and process data faster. Quantum learning can handle tricky classification problems, thanks to quantum parallelism and high-dimensional systems, which might go beyond what classical neural networks can do. QNNs are especially promising for data with built-in quantum features or where quantum entanglement helps solve the problem [8].

4.6.1.1 Model Designs

In the realm of quantum discriminative models, researchers have developed various quantum neural architectures. In general, variational quantum classifiers [68, 69] could employ parameterized quantum circuits for classification tasks. Subsequently, quantum convolutional neural networks [36] are designed for processing structured data. Hybrid quantum-classical architectures [70] are proposed to combine quantum layers with classical neural networks. Other notable works include the development of quantum versions of popular classical architectures like recurrent neural networks [71] and attention mechanisms [72]. Finally, Pérez-Salinas et al. [73] and Fan and Situ [74] have explored quantum re-uploading strategies for encoding classical data, achieving QML models with more expressive feature maps.

Besides designing these networks by hand, researchers are looking for ways to make QNNs simpler and faster. For example, quantum architecture search methods have been developed by Du et al. [75], Zhang et al. [76], Lu et al. [77], Linghu et al. [78] to automatically discover optimal quantum circuit designs with reduced gate complexity. Sim et al. [79] and Wang et al. [80] introduced quantum pruning techniques that systematically identify and remove redundant quantum gates while preserving the performance. In the realm of knowledge distillation, researchers have demonstrated how to transfer knowledge from the teacher model given as quantum [81] or classical [82] neural networks to more compact quantum circuit architectures that are more robust against quantum noises. These optimization approaches have collectively contributed to improving the practical performance of QNNs on real quantum devices, particularly in the NISQ era.

4.6.1.2 Theoretical Foundations

To find out where QNNs can really help, it is important to study how well they can learn. This usually comes down to three main ideas: flexibility (how well they model complex patterns), how easily a model can learn from data (trainability), and how well they work on new data (generalization). Researchers have dug deeper into each of these topics, which are described below.

Expressivity The expressivity of QNNs refers to their ability to represent complex functions or quantum states efficiently. Universal approximation theorems (UAT) incorporating data re-uploading strategies have been established by Pérez-Salinas et al. [73] firstly with subsequent works [83, 84] in various problem settings. Beyond the UAT, Sim et al. [85], Nakaji and Yamamoto [86], and Holmes et al. [87] analyze the expressivity of QNNs by investigating how well the parameterized quantum circuits used in QNNs can approximate the Haar distribution, a critical measure of expressive capacity in quantum systems. Moreover, Yu et al. [88] analyze the non-asymptotic error bounds of variational quantum circuits for approximating multivariate polynomials and smooth functions.

Trainability Trainability means how easily QNNs can learn from data. This depends on two things: how strong the learning signals ("slopes" or gradients) are and how quickly the model gets to a good solution.

For the first research line, McClean et al. [52] first found the phenomenon of vanishing gradients, dubbed as the barren plateau. That is, as quantum systems get bigger, the "slopes" used for learning almost disappear. This makes it very hard for QNNs to learn. Since then, a series of studies explored the cause of barren plateau, including global measurements [56], highly entanglement [89], quantum tensor networks [90], and quantum system noise [91].

To tackle this, researchers have tried smarter ways to pick starting parameters [59, 60], design new cost functions [56], and set up proper circuits [57]. Quantum-specific regularization techniques have also been developed to mitigate these effects [92].

Another research line looks at how fast QNNs can reach good solutions. Studies show that making QNNs large enough (overparameterized) helps them find good answers quickly. Kiani et al. [93] and Wiersema et al. [94] experimentally found that overparameterized QNNs embrace a benign landscape and permit fast convergence toward near optimal local minima. Later, theory caught up and explained why bigger QNNs learn so fast—even with complicated math behind the scenes. Specifically, Larocca et al. [51] and Anschuetz [95] utilized tools from dynamical Lie algebra and random matrix theory, respectively, to quantify the critical points in the optimization landscape of overparameterized QNNs. Moreover, You et al. [96] extended the classical convergence results of Xu et al. [97] to the quantum domain, proving that overparameterized QNNs achieve an exponential convergence rate. Additionally, Liu et al. [98] and Wang et al. [99] introduced the concept of the quantum neural tangent kernel (QNTK) to further demonstrate the exponential convergence rate of overparameterized QNNs. Besides the overparameterization theory, Du et al. [100, 101], Qi et al. [102], and Qian et al. [103] investigated the required conditions to ensure the convergence of QNNs toward local minima.

Generalization Another key question is do QNNs work well on new data? In particular, Abbas et al. [104] compared the generalization power of QNNs and classical learning models based on an information geometry metric. Caro et al. [105] and Du et al. [46] established generalization error bounds using covering numbers, revealing the impact of circuit structural factors—such as the number of gates and types of observables—on generalization ability. Similarly, Bu et al. [106] analyzed the generalization ability of QNNs from the perspective of quantum resource theory, emphasizing the role of quantum resources such as entanglement and magic in influencing generalization.

Some frameworks even point to situations where QNNs clearly beat classical models. In particular, Huang et al. [8, 107] provided insights into the conditions under which quantum models outperform their classical counterparts. Zhang et al. [108] investigate the training-dependent generalization abilities of QNNs, while Du et al. [109] study problem-dependent generalization, highlighting key factors that enable QNNs to achieve strong generalization performance. The generalization ability of QNNs under both the underparameterized and overparameterized regimes has been discussed by Qian et al. [110], Gil-Fuster et al. [111], and Kempkes et al. [112].

To get a big-picture view, the no-free-lunch theorem has been used to study QNNs. Poland et al. [113] explore the average performance of QNNs across all possible datasets and problems, providing a broader perspective on their generalization potential. Extending this work, Sharma et al. [114] and Wang et al. [115] adapt the no-free-lunch theorem to scenarios involving entangled data, demonstrating the potential benefits of entanglement in certain settings. Additionally, Wang et al. [116] establish a no-free-lunch framework for various learning protocols, considering different quantum resources used in these protocols.

Potential Advantages One major challenge in quantum learning is to find tasks where QNNs clearly beat classical models for solving "normal" (classical) problems. Some studies have shown QNNs can be better than classical models, but usually only on special, made-up tasks. For instance, polynomial advantages over commonly used classical models have been demonstrated through quantum contextuality in sequential learning [117, 118]. Additionally, quantum entanglement has been shown to reduce communication requirements in nonlocal machine learning tasks [119]. However, most QNNs—unless they are carefully designed—can be copied or simulated by classical computers. The review [120] presented strong evidence that commonly used models with provable absence of barren plateaus are also classically simulable. Concrete algorithms in this research line include Pauli path simulators [121–123], tensor-network simulators [124], and learning protocols [125–127].

4.6.1.3   Applications

QNNs for classifying data have found use in many areas. For example, researchers have tried quantum methods for image classification [128] and pattern recognition [129]. In quantum chemistry, QNNs help predict proton affinities [130] and aid in developing new catalysts [131]. In finance, they are used for tasks like classifying market trends [132] and detecting fraud [133]. In medicine, QNNs have supported drug discovery [134] and disease diagnosis [135, 136].

However, real-world tests show that today's QNNs still have important limitations [110, 137]. For example, one study compared eight popular QNNs to classical models on different datasets and found no clear performance advantage [137]. Still, QNNs can be very flexible on small, simple datasets. Sometimes, they need fewer parameters than classical networks to get similar results, since quantum systems can model more complicated patterns. Overall, QNNs for classification are still in the early stages. Better network designs, smarter training methods, and ways to reduce errors will be key for making quantum models truly useful.

4.6.2   Generative Learning with QNNs

QNNs for generating data are another exciting direction in quantum machine learning. These models use flexible quantum circuits to create data patterns that can be more complex than those made by classical models, especially in high-dimensional or quantum settings. For more details, see the survey by Tian et al. [67].

4.6.2.1   Model Designs

Researchers have designed many types of QNNs for generating data. For instance, quantum circuit Born machines (QCBMs) [45] use quantum circuits to produce random patterns. Quantum generative adversarial networks (QGANs) [41] bring the "generator vs. discriminator" idea into quantum computing. Quantum Boltzmann machines [138] use quantum systems to sample from complicated distributions. Quantum autoencoders [139] help with tasks like compressing or reconstructing quantum states. More recently, quantum diffusion models [140, 141] have been introduced to create sets of quantum states or even classical images. These examples show that QNNs can tackle a wide variety of generative problems.

4.6.2.2 Theoretical Foundations

On the theory side, generative QNNs face many of the same learning challenges as QNNs for classification. Similar to QNNs for discriminative tasks, quantum generative models like QCBMs face the barren plateau issue with additional mechanisms from the Kullback-Leibler (KL) divergence loss function [142]. But in cases where data is scarce, QCBMs can sometimes learn faster than classical generative models [143]. Some studies show that quantum models can represent data patterns that classical models cannot and can learn or generate these patterns much faster [144, 145]. For example, even simple quantum extensions of common classical models like Bayesian networks can model more complex relationships. Other research has explored how well quantum generative models can generalize to new data, using measures like maximum mean discrepancy [146].

4.6.2.3 Applications

Generative QNNs are being tested in a variety of fields. In finance, they have been used to create synthetic data and model tricky financial patterns, sometimes beating classical models [43, 147]. In physics, they help with tasks like simulating quantum systems or reconstructing quantum states [148]. For image generation, quantum GANs can produce high-quality pictures [44]. In drug discovery, quantum generative models may help search through large chemical spaces more efficiently [149]. These examples show the wide reach of QNNs in generative tasks.

References

1. LeCun, Y., Bengio, Y., & Hinton, G. (2015). Deep learning. *Nature, 521*(7553), 436–444.
2. Voulodimos, A., Doulamis, N., Doulamis, A., & Protopapadakis, E. (2018). Deep learning for computer vision: A brief review. *Computational Intelligence and Neuroscience, 2018*(1), 7068349.
3. Otter, D. W., Medina, J. R., & Kalita, J. K. (2020). A survey of the usages of deep learning for natural language processing. *IEEE Transactions on Neural Networks and Learning Systems, 32*(2), 604–624.
4. Hoffmann, J., Borgeaud, S., Mensch, A., Buchatskaya, E., Cai, T., Rutherford, E., de Las Casas, D., Hendricks, L. A., Welbl, J., Clark, A., et al. (2022). Training compute-optimal large language models. In *Proceedings of the 36th International Conference on Neural Information Processing Systems* (pp. 30016–30030).
5. de Vries, A. (2023). The growing energy footprint of artificial intelligence. *Joule, 7*(10), 2191–2194.
6. Jeswal, S. K., & Chakraverty, S. (2019). Recent developments and applications in quantum neural network: A review. *Archives of Computational Methods in Engineering, 26*(4), 793–807.
7. Liu, J., Liu, M., Liu, J.-P., Ye, Z., Wang, Y., Alexeev, Y., Eisert, J., & Jiang, L. (2024). Towards provably efficient quantum algorithms for large-scale machine-learning models. *Nature Communications, 15*(1), 434.

8. Huang, H.-Y., Broughton, M., Cotler, J., Chen, S., Li, J., Mohseni, M., Neven, H., Babbush, R., Kueng, R., Preskill, J., et al. (2022). Quantum advantage in learning from experiments. *Science, 376*(6598), 1182–1186.

9. Peters, E., Caldeira, J., Ho, A., Leichenauer, S., Mohseni, M., Neven, H., Spentzouris, P., Strain, D., & Perdue, G. N. (2021). Machine learning of high dimensional data on a noisy quantum processor. *NPJ Quantum Information, 7*(1), 161.

10. Acharya, R., Aghababaie-Beni, L., Aleiner, I., Andersen, T. I., Ansmann, M., Arute, F., Arya, K., Asfaw, A., Astrakhantsev, N., Atalaya, J. et al. (2024). Quantum error correction below the surface code threshold. arXiv preprint arXiv:2408.13687.

11. Gardas, B., Rams, M. M., & Dziarmaga, J. (2018). Quantum neural networks to simulate many-body quantum systems. *Physical Review B, 98*(18), 184304.

12. Cao, Y., Romero, J., Olson, J. P., Degroote, M., Johnson, P. D., Kieferová, M., Kivlichan, I. D., Menke, T., Peropadre, B., Sawaya, N. P. D., et al. (2019). Quantum chemistry in the age of quantum computing. *Chemical Reviews, 119*(19), 10856–10915.

13. Dua, D., & Graff, C. (2017). UCI machine learning repository. http://archive.ics.uci.edu/ml

14. LeCun, Y., Bottou, L., Bengio, Y., & Haffner, P. (1998). Gradient-based learning applied to document recognition. *Proceedings of the IEEE, 86*(11), 2278–2324.

15. McCulloch, W. S., & Pitts, W. (1943). A logical calculus of the ideas immanent in nervous activity. *The Bulletin of Mathematical Biophysics, 5*, 115–133.

16. Gardner, M. W., & Dorling, S. R. (1998). Artificial neural networks (the multilayer perceptron)—a review of applications in the atmospheric sciences. *Atmospheric Environment, 32*(14–15), 2627–2636.

17. Vaswani, A. (2017). Attention is all you need. *Advances in Neural Information Processing Systems, 30*, 5998–6008.

18. Hornik, K. (1993). Some new results on neural network approximation. *Neural Networks, 6*(8), 1069–1072.

19. Amari, S. (1993). Backpropagation and stochastic gradient descent method. *Neurocomputing, 5*(4–5), 185–196.

20. LeCun, Y., Boser, B., Denker, J., Henderson, D., Howard, R., Hubbard, W., & Jackel, L. (1989). Handwritten digit recognition with a back-propagation network. *Advances in Neural Information Processing Systems, 2*, 396–404.

21. He, K., Zhang, X., Ren, S., & Sun, J. (2016). Deep residual learning for image recognition. In *Proceedings of the IEEE Conference on Computer Vision and Pattern Recognition* (pp. 770–778).

22. Novikoff, A. B. J. (1962). On convergence proofs on perceptrons. In *Proceedings of the Symposium on the Mathematical Theory of Automata*, New York (Vol. 12, pp. 615–622).

23. Rosenblatt, F. (1958). The perceptron: A probabilistic model for information storage and organization in the brain. *Psychological Review, 65*(6), 386.

24. LeCun, Y., Touresky, D., Hinton, G., & Sejnowski, T. (1988). A theoretical framework for back-propagation. In *Proceedings of the 1988 Connectionist Models Summer School* (Vol. 1, pp. 21–28).

25. Hornik, K., Stinchcombe, M., & White, H. (1989). Multilayer feedforward networks are universal approximators. *Neural Networks, 2*(5), 359–366.

26. Caruana, R., Lawrence, S., & Giles, C. L. (2000). Overfitting in neural nets: Backpropagation, conjugate gradient, & early stopping. *Advances in Neural Information Processing Systems, 13*, 402–408.

27. Srivastava, N., Hinton, G., Krizhevsky, A., Sutskever, I., & Salakhutdinov, R. (2014). Dropout: A simple way to prevent neural networks from overfitting. *Journal of Machine Learning Research, 15*(1), 1929–1958.

28. Krogh, A., & Hertz, J. A. (1991). A simple weight decay can improve generalization. *Advances in Neural Information Processing Systems, 4*, 950–957.

29. Kapoor, A., Wiebe, N., & Svore, K. M. (2016). Quantum perceptron models. *Advances in Neural Information Processing Systems, 29*, 3999–4007.

30. Grover, L. K. (1996). A fast quantum mechanical algorithm for database search. In *Proceedings of the Twenty-Eighth Annual ACM Symposium on Theory of Computing* (pp. 212–219).

31. Arute, F., Arya, K., Babbush, R., Bacon, D., Bardin, J. C., Barends, R., Biswas, R., Boixo, S., Brandao, F.G. S. L., Buell, D. A., et al. (2019). Quantum supremacy using a programmable superconducting processor. *Nature, 574*(7779), 505–510.

32. AbuGhanem, M. (2024). Ibm quantum computers: Evolution, performance, and future directions. arXiv preprint arXiv:2410.00916.

33. Gao, D., Fan, D., Zha, C., Bei, J., Cai, G., Cai, J., Cao, S., Zeng, X., Chen, F., Chen, J., et al. (2024). Establishing a new benchmark in quantum computational advantage with 105-qubit zuchongzhi 3.0 processor. arXiv preprint arXiv:2412.11924.

34. Wang, Y., & Liu, J. (2024). A comprehensive review of quantum machine learning: From nisq to fault tolerance. *Reports on Progress in Physics, 87*(11), 116402.

35. Kandala, A., Mezzacapo, A., Temme, K., Takita, M., Brink, M., Chow, J. M., & Gambetta, J. M. (2017). Hardware-efficient variational quantum eigensolver for small molecules and quantum magnets. *Nature, 549*(7671), 242–246.

36. Cong, I., Choi, S., & Lukin, M. D. (2019). Quantum convolutional neural networks. *Nature Physics, 15*(12), 1273–1278.

37. Peruzzo, A., McClean, J., Shadbolt, P., Yung, M.-H., Zhou, X.-Q., Love, P. J., Aspuru-Guzik, A., & O'brien, J. L. (2014). A variational eigenvalue solver on a photonic quantum processor. *Nature Communications, 5*(1), 4213.

38. Crooks, G. E. (2019). Gradients of parameterized quantum gates using the parameter-shift rule and gate decomposition. arXiv preprint arXiv:1905.13311.

39. Duchi, J., Hazan, E., & Singer, Y. (2011). Adaptive subgradient methods for online learning and stochastic optimization. *Journal of Machine Learning Research, 12*(7), 2121–2159.

40. Kingma, D. P. (2014). Adam: A method for stochastic optimization. arXiv preprint arXiv:1412.6980.

41. Lloyd, S., & Weedbrook, C. (2018). Quantum generative adversarial learning. *Physical Review Letters, 121*(4), 040502.

42. Bravyi, S., Gosset, D., & König, R. (2018). Quantum advantage with shallow circuits. *Science, 362*(6412), 308–311.

43. Zhu, E. Y., Johri, S., Bacon, D., Esencan, M., Kim, J., Muir, M., Murgai, N., Nguyen, J., Pisenti, N., Schouela, A., et al. (2022). Generative quantum learning of joint probability distribution functions. *Physical Review Research, 4*(4), 043092.

44. Huang, H.-L., Du, Y., Gong, M., Zhao, Y., Wu, Y., Wang, C., Li, S., Liang, F., Lin, J., Xu, Y., et al. (2021a). Experimental quantum generative adversarial networks for image generation. *Physical Review Applied, 16*(2), 024051.

45. Benedetti, M., Garcia-Pintos, D., Perdomo, O., Leyton-Ortega, V., Nam, Y., & Perdomo-Ortiz, A. (2019a). A generative modeling approach for benchmarking and training shallow quantum circuits. *npj Quantum Information, 5*(1), 45.

46. Du, Y., Tu, Z., Yuan, X., & Tao, D. (2022a). Efficient measure for the expressivity of variational quantum algorithms. *Physical Review Letters, 128*(8), 080506.

47. Barthel, T., & Lu, J. (2018). Fundamental limitations for measurements in quantum many-body systems. *Physical Review Letters, 121*(8), 080406.

48. Kakade, S. M., Sridharan, K., & Tewari, A. (2008). On the complexity of linear prediction: Risk bounds, margin bounds, and regularization. *Advances in Neural Information Processing Systems, 21*, 793–800.

49. Dudley, R. M. (1967). The sizes of compact subsets of hilbert space and continuity of gaussian processes. *Journal of Functional Analysis, 1*(3), 290–330.

50. Haussler, A., & Warmuth, M. (1987). Occam's razor. *Information Processing Letters, 24*, 377–380.

51. Larocca, M., Ju, N., García-Martín, D., Coles, P. J., & Cerezo, M. (2023). Theory of overparametrization in quantum neural networks. *Nature Computational Science, 3*(6), 542–551.

52. McClean, J. R., Boixo, S., Smelyanskiy, V. N., Babbush, R., & Neven, H. (2018). Barren plateaus in quantum neural network training landscapes. *Nature Communications, 9*(1), 4812.

53. Dankert, C., Cleve, R., Emerson, J., & Livine, E. (2009). Exact and approximate unitary 2-designs and their application to fidelity estimation. *Physical Review A—Atomic, Molecular, and Optical Physics, 80*(1), 012304.

54. Harrow, A. W., & Low, R. A. (2009). Random quantum circuits are approximate 2-designs. *Communications in Mathematical Physics, 291*, 257–302.

55. Haferkamp, J. (2022). Random quantum circuits are approximate unitary t-designs in depth $o(nt^{5+o(1)})$. *Quantum, 6*, 795.

56. Cerezo, M., Sone, A., Volkoff, T., Cincio, L., & Coles, P. J. (2021). Cost function dependent barren plateaus in shallow parametrized quantum circuits. *Nature Communications, 12*(1), 1791.

57. Pesah, A., Cerezo, M., Wang, S., Volkoff, T., Sornborger, A. T., & Coles, P. J. (2021). Absence of barren plateaus in quantum convolutional neural networks. *Physical Review X, 11*(4), 041011.

58. Zhang, K., Hsieh, M.-H., Liu, L., & Tao, D. (2021). Toward trainability of deep quantum neural networks. arXiv preprint arXiv:2112.15002.

59. Grant, E., Wossnig, L., Ostaszewski, M., & Benedetti, M. (2019). An initialization strategy for addressing barren plateaus in parametrized quantum circuits. *Quantum, 3*, 214.

60. Zhang, K., Liu, L., Hsieh, M.-H., & Tao, D. (2022a). Escaping from the barren plateau via gaussian initializations in deep variational quantum circuits. *Advances in Neural Information Processing Systems, 35*, 18612–18627.

61. Skolik, A., McClean, J. R., Mohseni, M., Van Der Smagt, P., & Leib, M. (2021). Layerwise learning for quantum neural networks. *Quantum Machine Intelligence, 3*, 1–11.

62. Haug, T., & Kim, M. S. (2021). Optimal training of variational quantum algorithms without barren plateaus. arXiv preprint arXiv:2104.14543.

63. Ablayev, F., Ablayev, M., Huang, J. Z., Khadiev, K., Salikhova, N., & Wu, D. (2019). On quantum methods for machine learning problems part ii: Quantum classification algorithms. *Big Data Mining and Analytics, 3*(1), 56–67.

64. Li, W., & Deng, D.-L. (2022). Recent advances for quantum classifiers. *Science China Physics, Mechanics & Astronomy, 65*(2), 220301.

65. Massoli, F. V., Vadicamo, L., Amato, G., & Falchi, F. (2022). A leap among quantum computing and quantum neural networks: A survey. *ACM Computing Surveys, 55*(5), 1–37.

66. Li, W., Lu, Z., & Deng, D.-L. (2022). Quantum neural network classifiers: A tutorial. *SciPost Physics Lecture Notes, 61*, 1–28.

67. Tian, J., Sun, X., Du, Y., Zhao, S., Liu, Q., Zhang, K., Yi, W., Huang, W., Wang, C., Wu, X., et al. (2023). Recent advances for quantum neural networks in generative learning. *IEEE Transactions on Pattern Analysis and Machine Intelligence, 45*(10), 12321–12340.

68. Havlicek, V., Corcoles, A. D., Temme, K., Harrow, A. W., Kandala, A., Chow, J. M., & Gambetta, J. M. (2019). Supervised learning with quantum-enhanced feature spaces. *Nature, 567*(7747), 209–212.

69. Mitarai, K., Negoro, M., Kitagawa, M., & Fujii, K. (2018). Quantum circuit learning. *Physical Review A, 98*(3), 032309.

70. Arthur, D., et al. (2022). A hybrid quantum-classical neural network architecture for binary classification. arXiv preprint arXiv:2201.01820.

71. Bausch, J. (2020). Recurrent quantum neural networks. *Advances in Neural Information Processing Systems, 33*, 1368–1379.

72. Shi, J., Zhao, R.-X., Wang, W., Zhang, S., & Li, X. (2024). QSAN: A near-term achievable quantum self-attention network. *IEEE Transactions on Neural Networks and Learning Systems, 36*(8), 13995–14008.

73. Pérez-Salinas, A., Cervera-Lierta, A., Gil-Fuster, E., & Latorre, J. (2020). Data re-uploading for a universal quantum classifier. *Quantum, 4*, 226.

74. Fan, L., & Situ, H. (2022). Compact data encoding for data re-uploading quantum classifier. *Quantum Information Processing, 21*(3), 87.

75. Du, Y., Huang, T., You, S., Hsieh, M.-H., & Tao, D. (2022b). Quantum circuit architecture search for variational quantum algorithms. *npj Quantum Information, 8*(1), 62.

76. Zhang, S.-X., Hsieh, C.-Y., Zhang, S., & Yao, H. (2022b). Differentiable quantum architecture search. *Quantum Science and Technology, 7*(4), 045023.

77. Lu, Z., Shen, P.-X., & Deng, D.-L. (2021). Markovian quantum neuroevolution for machine learning. *Physical Review Applied, 16*(4), 044039.

78. Linghu, K., Qian, Y., Wang, R., Hu, M.-J., Li, Z., Li, X., Xu, H., Zhang, J., Ma, T., Zhao, P., et al. Quantum circuit architecture search on a superconducting processor. *Entropy, 26*(12), 1025 (2024).

79. Sim, S., Romero, J., Gonthier, J. F. & Kunitsa, A. A. (2021). Adaptive pruning-based optimization of parameterized quantum circuits. *Quantum Science and Technology, 6*(2), 025019.

80. Wang, X., Liu, J., Liu, T., Luo, Y., Du, Y., & Tao, D. (2022). Symmetric pruning in quantum neural networks. arXiv preprint arXiv:2208.14057.

81. Alam, M., Kundu, S., & Ghosh, S. (2023). Knowledge distillation in quantum neural network using approximate synthesis. In *Proceedings of the 28th Asia and South Pacific Design Automation Conference* (pp. 639–644).

82. Li, M., Fan, L., Cummings, A., Zhang, X., Pan, M., & Han, Z. (2024). Hybrid quantum classical machine learning with knowledge distillation. In *ICC 2024-IEEE International Conference on Communications* (pp. 1139–1144). IEEE.

83. Schuld, M., Sweke, R., & Meyer, J. J. (2021). Effect of data encoding on the expressive power of variational quantum-machine-learning models. *Physical Review A, 103*(3), 032430.

84. Yu, Z., Yao, H., Li, M., & Wang, X. (2022). Power and limitations of single-qubit native quantum neural networks. In S. Koyejo, S. Mohamed, A. Agarwal, D. Belgrave, K. Cho, & A. Oh (Eds.), *Advances in neural information processing systems* (Vol. 35, pp. 27810–27823). Curran Associates, Inc.

85. Sim, S., Johnson, P. D., & Aspuru-Guzik, A. (2019). Expressibility and entangling capability of parameterized quantum circuits for hybrid quantum-classical algorithms. *Advanced Quantum Technologies, 2*(12), 1900070.

86. Nakaji, K., & Yamamoto, N. (2021). Expressibility of the alternating layered ansatz for quantum computation. *Quantum, 5*, 434.

87. Holmes, Z., Sharma, K., Cerezo, M., & Coles, P. J. (2022). Connecting ansatz expressibility to gradient magnitudes and barren plateaus. *PRX Quantum, 3*(1), 010313.

88. Yu, Z., Chen, Q., Jiao, Y., Li, Y., Lu, X., Wang, X., & Yang, J. (2025). Non-asymptotic approximation error bounds of parameterized quantum circuits. *Advances in Neural Information Processing Systems, 37*, 99089–99127.

89. Ortiz Marrero, C., Kieferová, M., & Wiebe, N. (2021). Entanglement-induced barren plateaus. *PRX Quantum, 2*(4), 040316.

90. Martin, E. C., Plekhanov, K., & Lubasch, M. (2023). Barren plateaus in quantum tensor network optimization. *Quantum, 7*, 974.

91. Wang, S., Fontana, E., Cerezo, M., Sharma, K., Sone, A., Cincio, L., & Coles, P. J. (2021). Noise-induced barren plateaus in variational quantum algorithms. *Nature Communications, 12*(1), 6961.

92. Larocca, M., Czarnik, P., Sharma, K., Muraleedharan, G., Coles, P. J., & Cerezo, M. (2022). Diagnosing barren plateaus with tools from quantum optimal control. *Quantum, 6*, 824.

93. Kiani, B. T., Lloyd, S., & Maity, R. (2020). Learning unitaries by gradient descent. arXiv preprint arXiv:2001.11897.

94. Wiersema, R., Zhou, C., de Sereville, Y., Carrasquilla, J. F., Kim, Y. B., & Yuen, H. (2020). Exploring entanglement and optimization within the hamiltonian variational ansatz. *PRX Quantum, 1*(2), 020319.

95. Anschuetz, E. R. (2021). Critical points in quantum generative models. arXiv preprint arXiv:2109.06957.

96. You, X., Chakrabarti, S., & Wu, X. (2022). A convergence theory for over-parameterized variational quantum eigensolvers. arXiv preprint arXiv:2205.12481.

97. Xu, Z., Cao, X., & Gao, X. (2018). Convergence analysis of gradient descent for eigenvector computation. In *International Joint Conferences on Artificial Intelligence*.

98. Liu, J., Najafi, K., Sharma, K., Tacchino, F., Jiang, L., & Mezzacapo, A. (2023). Analytic theory for the dynamics of wide quantum neural networks. *Physical Review Letters, 130*(15), 150601.

99. Wang, X., Liu, J., Liu, T., Luo, Y., Du, Y., & Tao, D. (2023). Symmetric pruning in quantum neural networks. In *The Eleventh International Conference on Learning Representations*. https://openreview.net/forum?id=K96AogLDT2K

100. Du, Y., Hsieh, M.-H., Liu, T., You, S., & Tao, D. (2021). Learnability of quantum neural networks. *PRX Quantum, 2*(4), 040337.

101. Du, Y., Qian, Y., Wu, X., & Tao, D. (2022c). A distributed learning scheme for variational quantum algorithms. *IEEE Transactions on Quantum Engineering, 3*, 1–16.

102. Qi, J., Yang, C.-H. H., Chen, P.-Y. & Hsieh, M.-H. (2023). Theoretical error performance analysis for variational quantum circuit based functional regression. *npj Quantum Information, 9*(1), 4.

103. Qian, Y., Du, Y., & Tao, D. (2024). Shuffle-qudio: Accelerate distributed vqe with trainability enhancement and measurement reduction. *Quantum Machine Intelligence, 6*(1), 1–22.

104. Abbas, A., Sutter, D., Zoufal, C., Lucchi, A., Figalli, A., & Woerner, S. (2021). The power of quantum neural networks. *Nature Computational Science, 1*(6), 403–409.

105. Caro, M. C., Huang, H.-Y., Cerezo, M., Sharma, K., Sornborger, A., Cincio, L., & Coles, P. J. (2022). Generalization in quantum machine learning from few training data. *Nature Communications, 13*(1), 4919.

106. Bu, K., Koh, D. E., Li, L., Luo, Q., & Zhang, Y. (2022). Statistical complexity of quantum circuits. *Physical Review A, 105*(6), 062431.

107. Huang, H.-Y., Kueng, R., & Preskill, J. (2021b). Information-theoretic bounds on quantum advantage in machine learning. *Physical Review Letters, 126*(19), 190505.

108. Zhang, K., Liu, J., Liu, L., Jiang, L., Hsieh, M.-H., & Tao, D. (2024a). The curse of random quantum data. arXiv preprint arXiv:2408.09937.

109. Du, Y., Yang, Y., Tao, D., & Hsieh, M.-H. (2023). Problem-dependent power of quantum neural networks on multiclass classification. *Physical Review Letters, 131*(14), 140601.

110. Qian, Y., Wang, X., Du, Y., Wu, X., & Tao, D. (2022). The dilemma of quantum neural networks. *IEEE Transactions on Neural Networks and Learning Systems, 35*(4), 5603–5615.

111. Gil-Fuster, E., Eisert, J., & Bravo-Prieto, C. (2024). Understanding quantum machine learning also requires rethinking generalization. *Nature Communications, 15*(1), 2277.

112. Kempkes, M., Ijaz, A., Gil-Fuster, E., Bravo-Prieto, C., Spiegelberg, J., van Nieuwenburg, E., & Dunjko, V. (2025). Double descent in quantum machine learning. arXiv preprint arXiv:2501.10077.

113. Poland, K., Beer, K., & Osborne, T. J. (2020). No free lunch for quantum machine learning. arXiv preprint arXiv:2003.14103.

114. Sharma, K., Cerezo, M., Holmes, Z., Cincio, L., Sornborger, A., & Coles, P. J. (2022). Reformulation of the no-free-lunch theorem for entangled datasets. *Physical Review Letters, 128*(7), 070501.

115. Wang, X., Du, Y., Tu, Z., Luo, Y., Yuan, X., & Tao, D. (2024a). Transition role of entangled data in quantum machine learning. *Nature Communications, 15*(1), 3716.

116. Wang, X., Du, Y., Liu, K., Luo, Y., Du, B., & Tao, D. (2024b). Separable power of classical and quantum learning protocols through the lens of no-free-lunch theorem. arXiv preprint arXiv:2405.07226.

117. Anschuetz, E. R., Hu, H.-Y., Huang, J.-L., & Gao, X. (2023). Interpretable quantum advantage in neural sequence learning. *PRX Quantum, 4*(2), 020338.

118. Anschuetz, E. R., & Gao, X. (2024). Arbitrary polynomial separations in trainable quantum machine learning. arXiv preprint arXiv:2402.08606.

119. Zhao, H., & Deng, D.-L. (2024). Entanglement-induced provable and robust quantum learning advantages. arXiv preprint arXiv:2410.03094.

120. Cerezo, M., Larocca, M., García-Martín, D., Diaz, N. L., Braccia, P., Fontana, E., Rudolph, M. S., Bermejo, P., Ijaz, A., Thanasilp, S., et al. (2023). Does provable absence of barren plateaus imply classical simulability? or, why we need to rethink variational quantum computing. arXiv preprint arXiv:2312.09121.

121. Bermejo, P., Braccia, P., Rudolph, M. S., Holmes, Z., Cincio, L., & Cerezo, M. (2024). Quantum convolutional neural networks are (effectively) classically simulable. arXiv preprint arXiv:2408.12739.

122. Angrisani, A., Schmidhuber, A., Rudolph, M. S., Cerezo, M., Holmes, Z., & Huang, H.-Y. (2024). Classically estimating observables of noiseless quantum circuits. arXiv preprint arXiv:2409.01706.

123. Lerch, S., Puig, R., Rudolph, M. S., Angrisani, A., Jones, T., Cerezo, M., Thanasilp, S., & Holmes, Z. (2024). Efficient quantum-enhanced classical simulation for patches of quantum landscapes. arXiv preprint arXiv:2411.19896.

124. Shin, S., Teo, Y. S., & Jeong, H. (2024). Dequantizing quantum machine learning models using tensor networks. *Physical Review Research, 6*(2), 023218.

125. Landman, J., Thabet, S., Dalyac, C., Mhiri, H., & Kashefi, E. (2022). Classically approximating variational quantum machine learning with random fourier features. arXiv preprint arXiv:2210.13200.

126. Schreiber, F. J., Eisert, J., & Meyer, J. J. (2023). Classical surrogates for quantum learning models. *Physical Review Letters, 131*(10), 100803.

127. Du, Y., Hsieh, M.-H., & Tao, D. (2024). Efficient learning for linear properties of bounded-gate quantum circuits. arXiv preprint arXiv:2408.12199.

128. Henderson, M., Shakya, S., Pradhan, S., & Cook, T. (2020). Quanvolutional neural networks: Powering image recognition with quantum circuits. *Quantum Machine Intelligence, 2*(1), 2.

129. Alrikabi, H. T. S., Aljazaery, I. A., Qateef, J. S., Alaidi, A. H. M., & Roa'a, M. (2022). Face patterns analysis and recognition system based on quantum neural network QNN. *International Journal of Interactive Mobile Technologies, 16*(8), 35–48.

130. Jin, H., & Merz Jr., K. M. (2024). Integrating machine learning and quantum circuits for proton affinity predictions. arXiv preprint arXiv:2411.17856.

131. Roh, J., Oh, S., Lee, D., Joo, C., Park, J., Moon, I., Ro, I., & Kim, J. (2024). Hybrid quantum neural network model with catalyst experimental validation: Application for the dry reforming of methane. *ACS Sustainable Chemistry & Engineering, 12*(10), 4121–4131.

132. Li, Y., Wang, Z., Han, R., Shi, S., Li, J., Shang, R., Zheng, H., Zhong, G., & Gu, Y. (2023). Quantum recurrent neural networks for sequential learning. *Neural Networks, 166*, 148–161.

133. Innan, N., Sawaika, A., Dhor, A., Dutta, S., Thota, S., Gokal, H., Patel, N., Khan, M. A., Theodonis, I., & Bennai, M. (2024). Financial fraud detection using quantum graph neural networks. *Quantum Machine Intelligence, 6*(1), 7.

134. Batra, K., Zorn, K. M., Foil, D. H., Minerali, E., Gawriljuk, V. O., Lane, T. R., & Ekins, S. (2021). Quantum machine learning algorithms for drug discovery applications. *Journal of Chemical Information and Modeling, 61*(6), 2641–2647.

135. Benedetti, M., Coyle, B., Fiorentini, M., Lubasch, M., & Rosenkranz, M. (2021). Variational inference with a quantum computer. *Physical Review Applied, 16*(4), 044057.

136. Enad, H. G., Mohammed, M. A., et al. (2023). A review on artificial intelligence and quantum machine learning for heart disease diagnosis: Current techniques, challenges and issues, recent developments, & future directions. *Fusion: Practice and Applications (FPA), 11*(1), 08–25.

137. Bowles, J., Ahmed, S., & Schuld, M. (2024). Better than classical? the subtle art of benchmarking quantum machine learning models. arXiv preprint arXiv:2403.07059.

138. Amin, M. H., Andriyash, E., Rolfe, J., Kulchytskyy, B., & Melko, R. (2018). Quantum boltzmann machine. *Physical Review X, 8*(2), 021050.

139. Romero, J., Olson, J. P., & Aspuru-Guzik, A. (2017). Quantum autoencoders for efficient compression of quantum data. *Quantum Science and Technology, 2*(4), 045001.

140. Zhang, B., Xu, P., Chen, X., & Zhuang, Q. (2024b). Generative quantum machine learning via denoising diffusion probabilistic models. *Physical Review Letters, 132*(10), 100602.

141. Kölle, M., Stenzel, G., Stein, J., Zielinski, S., Ommer, B., & Linnhoff-Popien, C. (2024). Quantum denoising diffusion models. arXiv preprint arXiv:2401.07049.
142. Rudolph, M. S., Lerch, S., Thanasilp, S., Kiss, O., Shaya, O., Vallecorsa, S., Grossi, M., & Holmes, Z. (2024). Trainability barriers and opportunities in quantum generative modeling. *npj Quantum Information, 10*(1), 116.
143. Hibat-Allah, M., Mauri, M., Carrasquilla, J., & Perdomo-Ortiz, A. (2024). A framework for demonstrating practical quantum advantage: Comparing quantum against classical generative models. *Communications Physics, 7*(1), 68.
144. Gao, X., Zhang, Z.-Y., & Duan, L.-M. (2018). A quantum machine learning algorithm based on generative models. *Science Advances, 4*(12), eaat9004.
145. Gao, X., Anschuetz, E. R., Wang, S.-T., Cirac, J. I., & Lukin, M. D. (2022). Enhancing generative models via quantum correlations. *Physical Review X, 12*(2), 021037.
146. Du, Y., Tu, Z., Wu, B., Yuan, X., & Tao, D. (2022d). Power of quantum generative learning. arXiv preprint arXiv:2205.04730.
147. Alcazar, J., Leyton-Ortega, V., & Perdomo-Ortiz, A. (2020). Classical versus quantum models in machine learning: Insights from a finance application. *Machine Learning: Science and Technology, 1*(3), 035003.
148. Benedetti, M., Grant, E., Wossnig, L., & Severini, S. (2019b). Adversarial quantum circuit learning for pure state approximation. *New Journal of Physics, 21*(4), 043023.
149. Li, J., Topaloglu, R. O., & Ghosh, S. (2021). Quantum generative models for small molecule drug discovery. *IEEE Transactions on Quantum Engineering, 2*, 1–8.

Chapter 5
Quantum Transformer

Abstract This chapter provides a comprehensive introduction to the quantum transformer algorithm. In Sect. 5.1 we first described what is the transformer architecture, with detailed explanation about its key subroutines. We also briefly mention the optimization and training. In Sect. 5.2, we provide a guide about designing each quantum subroutine, including quantum self-attention, quantum residual connection with layer norm, and quantum feed-forward neural networks, based on the quantum linear algebra. We further mention various numerical studies on the open-source large language models and provide a detailed discussion about the potential of quantum advantage in Sect. 5.3. Some basic codes are provided in Sect. 5.4. Finally in Sect. 5.5, we provide a bibliographic remark for readers who are interested to explore.

Transformers, introduced by Vaswani [1], have become one of the most important and widely adopted deep learning architectures in modern AI. Transformers were first developed to improve previous architectures for natural language processing on the ability to handle long-range dependencies and capture intricate relationships in data. Unlike previous sequential models, such as recurrent neural networks, which process information in a step-by-step manner, transformers use a mechanism called *self-attention* to capture correlations among all elements in a sequence simultaneously. This parallel processing capability significantly reduces training time and improves learning performance.

Despite its many advantages, the transformer architecture has several drawbacks, particularly the required computational resources. As discussed in previous chapters, quantum computing provides unique advantages over classical computing in certain applications by leveraging quantum phenomena such as superposition, entanglement, and interference. These capabilities have inspired researchers to explore whether integrating quantum computing with Transformers could lead to superior performance compared to their classical counterparts in specific tasks.

To address this question, in this chapter, the mechanism of Transformers is introduced in Sect. 5.1. Then, the construction of a quantum Transformer on a fault-tolerant quantum computer is illustrated in Sect. 5.2. The runtime of quantum

Y. Du et al., *A Gentle Introduction to Quantum Machine Learning*,
https://doi.org/10.1007/978-981-95-1284-3_5

transformers combined with numerical observations is also analyzed in Sect. 5.3, demonstrating a quadratic speedup over the classical counterpart.

5.1 Classical Transformer

The transformer architecture is designed to predict the next *token* (formally present in Sect. 5.1.1) in a sequence by leveraging sophisticated neural network components. Its modular design—including residual connections, layer normalization, and feed-forward networks (FFNs) as introduced in Sect. 4.1—makes it highly scalable and customizable. This versatility has enabled its successful application in large-scale foundation models across diverse domains. Notable examples include natural language processing, computer vision, reinforcement learning, robotics, and beyond.

The full architecture of Transformer is illustrated in Fig. 5.1. In particular, the encoder processes the source sequence through multiple layers of multiheaded self-attention and feed-forward networks, augmented with residual connections,

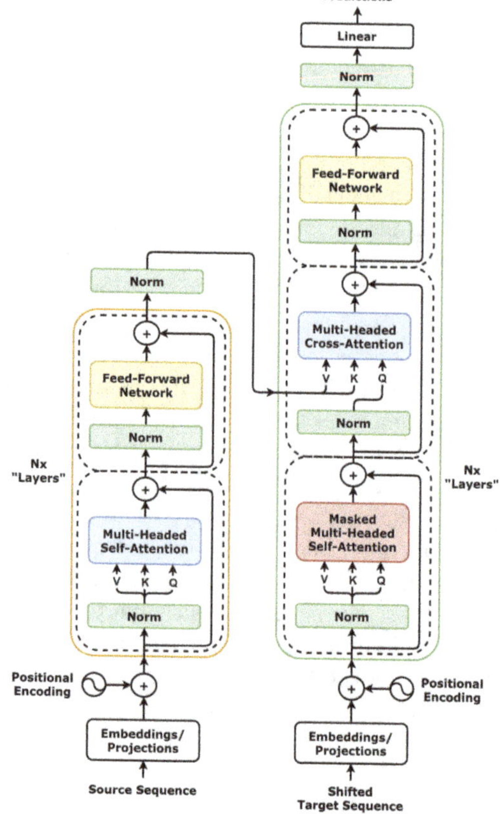

Fig. 5.1 A standard transformer architecture, showing an encoder on the left and a decoder on the right. (Image by Daniel Voigt Godoy under CC BY)

layer normalization, and positional encodings. The decoder incorporates masked multiheaded self-attention to process the target sequence and multiheaded cross-attention to integrate information from the encoder. The output is passed through a feed-forward network and a final linear layer to generate predictions. Note that while the original paper by Vaswani [1] introduced both encoder and decoder components, contemporary large language models primarily adopt *decoder-only architectures*, which have demonstrated superior practical performance. Therefore, in the remainder of this section, the implementation of each building block is detailed, and the optimization of decoder-only Transformer architectures is discussed. To deepen the understanding, a toy example of a classical Transformer with the code demonstration is provided in Sect. 5.4.

5.1.1 Tokenization and Embedding

To handle sequential data, such as natural language, Transformers employ *tokenization* to convert it into discrete units. This preprocessing step makes the data compatible with computational models and optimizes it for parallel processing, particularly on GPUs. More concisely, Tokenization breaks a sentence into smaller pieces called *tokens*, which could be words, subwords, or even characters, depending on the tokenization strategy. For example, the sentence "Transformers are amazing!" might become tokens like ("Transform," "ers," "are," "amazing," "!") if subwords are used. Modern tokenization methods [2–4] enable sophisticated mapping of complex inputs into token spaces.

For Transformer, tokens are mapped to high-dimensional real vector representations via *embedding* [1], as highlighted by the solid box "Embeddings/Projections" in Fig. 5.1. Let d_{token} denote the dictionary's token count and d_{model} represent the embedding vector dimension. The set containing all token embedding vectors in the dictionary is defined as

$$\mathcal{W} := \left\{ \mathcal{W}_j \in \mathbb{R}^{d_{\text{model}}} : \mathcal{W}_j \text{ is the embedding of token } j \in [d_{\text{token}}] \right\}.$$

An ℓ-length sentence is represented as a sequence of vectors $\{S_j\}_{j=1}^{\ell}$, where $S_j \in \mathcal{W}$. Mathematically, this sequence can be interpreted as a real matrix $S \in \mathbb{R}^{\ell \times d_{\text{model}}}$ whose j-th row S_j representing the j-th token.

5.1.2 Self-Attention

Self-attention is a core building block of the transformer architecture, which captures intrinsic correlations among tokens. By allowing each token in a sequence to attend to every other token, Transformer generates attention matrices via the inner-product operations. This operation encodes complex inter-token relationships into a transformative vector representation, colloquially termed "scaled dot-product attention." The generated attention matrices highlight how relevant each part of the input is to every other part. This allows Transformers to handle contextual dependencies across a variety of data structures.

The self-attention mechanism, as highlighted by the blue or red box in Fig. 5.1, involves three parameterized weight matrices: W_q, $W_k \in \mathbb{R}^{d_{\text{model}} \times d_k}$, and $W_v \in \mathbb{R}^{d_{\text{model}} \times d_v}$.

Remark

Following conventions [1], the notation d is used to specify d_{model}, d_k, and d_v in the rest of this chapter, i.e., $d := d_{\text{model}} = d_k = d_v$. This is a widely used setting in practice.

Given a sequence $S \in \mathbb{R}^{\ell \times d}$, the three new matrices after interacting it with three parameterized weight matrices W_q, W_k, W_v are defined, i.e.,

- Query matrix: $Q := S W_q$.
- Key matrix: $K := S W_k$.
- Value matrix: $V := S W_v$.

The attention block computes the matrix $G^{\text{soft}} \in \mathbb{R}^{\ell \times d}$ via

$$\text{Attention}(Q, K, V) = \text{softmax}\left(Q K^\top / \alpha_0\right) V =: G^{\text{soft}}, \tag{5.1}$$

where $\alpha_0 > 0$ is a scaling factor and $\text{softmax}(\cdot)$ is a row-wise nonlinear transformation such that $\text{softmax}(z)_j := e^{z_j} / \left(\sum_{k \in [\ell]} e^{z_k}\right)$ for $z \in \mathbb{R}^\ell$ and $j \in [\ell]$. The row-wise softmax application ensures the controlled attention distribution. The scaling factor $\alpha_0 = \sqrt{d}$ empirically prevents excessive value amplification, particularly when input matrix rows have zero mean and unit standard deviation.

For decoder-only architectures, *masked* self-attention is employed, strategically hiding tokens subsequent to the current query token, i.e.,

$$M_{jk} = \begin{cases} 0 & k \le j, \\ -\infty & k > j. \end{cases} \tag{5.2}$$

Conceptually, the mask M is applied to the scaled dot product $Q K^\top / \alpha_0$ in Eq. (5.1) before the softmax operation. Specifically, the matrix in the softmax operation is modified as $Q K^\top / \alpha_0 + M$.

Another crucial technique in Transformer is the multihead attention, which further extends the self-attention mechanism by computing and concatenating multiple attention matrices. This operation enables parallel representation learning across different subspaces. In practice, the embedding dimensions are often much larger (e.g., $d = 512$ or $d = 768$), with multiple attention heads working simultaneously, each capturing distinct relationships between words in the sequence. This mechanism plays a crucial role in modern AI, as it allows words to dynamically interact with one another within the context of the sequence.

While multihead attention is a pivotal advancement in Transformer architectures, its details will not be explored here, as *the quantum Transformer implementations presented below primarily support single-head attention.* Nonetheless, these techniques remain essential in classical AI and are promising directions for future developments in quantum Transformers.

5.1.3 Residual Connection

Residual connections (the arrows bypassing the main components, such as the attention and feed-forward layers in Fig. 5.1), paired with layer normalization (green box in the figure), provide crucial architectural flexibility and robustness. By enabling direct information flow between layers, they mitigate challenges in training deep neural networks [5, 6].

For the j-th token in an ℓ-length sentence, the residual connection generates $G_j^{\text{soft}} + S_j \in \mathbb{R}^d$ for $\forall j \in [\ell]$, which is subsequently normalized to standardize the vector representation. Let

$$\bar{s}_j := \frac{1}{d} \left(\sum_{k=1}^{d} \left(G_{jk}^{\text{soft}} + S_{jk} \right), \dots, \sum_{k=1}^{d} \left(G_{jk}^{\text{soft}} + S_{jk} \right) \right) \in \mathbb{R}^d,$$

where $\varsigma := \sqrt{\frac{1}{d} \sum_{k=1}^{d} \left(\left(G_j^{\text{soft}} + S_j - \bar{s}_j \right)_k \right)^2}$. The complete residual connection with the *layer normalization* $\text{LN}(\cdot, \cdot)$ can be expressed as

$$\text{LN}_{\gamma, \beta} \left(G_j^{\text{soft}}, S_j \right) = \gamma \frac{G_j^{\text{soft}} + S_j - \bar{s}_j}{\varsigma} + \beta, \tag{5.3}$$

where γ and β denote the scale and bias parameters, respectively.

5.1.4 Feed-Forward Network

Recall the definitions of fully connected neural networks (FFN) in Sect. 4.1. Transformers employ a two-layer fully connected transformation (yellow box in Fig. 5.1) to proceed with the output of residual connection, i.e.,

$$\text{FFN}\left(\text{LN}(z_j, S_j) \right) = \sigma \left(\text{LN}\left(G_j^{\text{soft}}, S_j \right) M_1 + b_1 \right) M_2 + b_2, \tag{5.4}$$

where $M_1 \in \mathbb{R}^{d \times d'}$, $M_2 \in \mathbb{R}^{d' \times d}$ are linear transformation matrices and b_1, b_2 are vectors. In most practical cases, $d' = 4d$. Here, $\sigma(x)$ is an activation function, such as $\tanh(x)$ and $\text{ReLU}(x) = \max(0, x)$. Another activation function that has been widely used in Transformers is the Gaussian error linear unit function (GELU), i.e.,

$$\text{GELU}(x) = x \cdot \frac{1}{2} \left(1 + \text{erf}\left(\frac{x}{\sqrt{2}} \right) \right).$$

Single-Head and Single-Block Transformer

By integrating all ingredients introduced in Sect. 5.1.1, the explicit form of single-head and single-block Transformer is reached, i.e.,

$$\text{Transformer}(S, j) := \text{LN}(\text{FFN}(\text{LN}(\text{Attention}(S, j)))). \tag{5.5}$$

For the final output, i.e., to predict the next token, one can implement the softmax function to make the vector into a distribution $\Pr(\cdot | S_1, \ldots, S_{j-1})$, where the dimension is the size of the token dictionary d_{token}, and sample from this distribution.

In modern architectures, multiple computational blocks are applied iteratively. Similar to multihead attention, we will not explore this in detail here, as the quantum Transformer implementations introduced below primarily support single-head attention.

5.1.5 Optimization and Inference

Upon the architecture of Transformer, its *optimization* involves training the model to achieve high performance on a given task. When applied to language processing tasks, a feasible loss function of the Transformer is the cross entropy between the predicted probability distribution and the correct distribution. Mathematically, given a ℓ-length sequence $\{S_1, \ldots, S_\ell\}$ as input, the loss function can be written as

$$\mathcal{L} = -\frac{1}{\ell} \sum_{j=1}^{\ell} \Pr(S_j | S_1, \ldots, S_{j-1}), \tag{5.6}$$

where $\Pr(S_j | S_1, \ldots, S_{j-1})$ is the predicted probability of the correct S_j coming out of the softmax layer, based on the previous tokens.

This process of training Transformers can be achieved by using Adam optimizer [7]. Learning rate schedules, such as warm-up followed by decay, can provide stable and effective convergence.

After optimization, the trained Transformer can be used in *inference*, which refers to making predictions on new data. Given a new initial sequence $S' = \{S'_1, \ldots, S'_{j-1}\}$, it is fed to the trained Transformer to obtain the distribution $\Pr(S'_j | S'_1, \ldots, S'_{j-1})$. Then, a decoding strategy (e.g., greedy decoding or sampling) can be used to select the token based on the highest probability or a sampling approach.

Efficient inference is crucial for deploying models in real-world applications. Speed optimization techniques, such as quantization, reduce the precision of weights and activations to accelerate inference with minimal accuracy loss [8]. Pruning removes redundant weights or attention heads to reduce the model size and computational cost [9]. Batching and parallelism are also critical, with batched inference allowing the processing of multiple inputs simultaneously and GPU or TPU acceleration enabling parallel computations. Efficient attention at inference, such as caching key and value tensors, reduces redundant computations in autoregressive tasks like text generation [10].

The inference cost is up to ten times the training cost as large language models (LLMs) are trained once and applied millions of times [11, 12]. For this reason, in the next section, the harnessing of quantum computing to address this issue is explored, which is crucial from both scientific and societal perspectives.

5.2 Fault-Tolerant Quantum Transformer

In this section, an end-to-end transformer architecture implementable on a quantum device is presented, which includes all the key building blocks introduced in Sect. 5.1. Besides, the potential runtime speedups of this quantum model are discussed. In particular, here the focus is on the *inference process* in which a classical Transformer has already been trained and is queried to predict the single next token.

Recall that in Sect. 5.1, the three parameterized matrices in the self-attention mechanism are assumed to have the same size, i.e., $W_q, W_k, W_v \in \mathbb{R}^{d \times d}$. Besides, the input sequence S and the matrix returned by the attention block G^{soft} have the size $\ell \times d$. Here, it is further supposed that the length of the sentence and the dimension of hidden features exponentially scale with 2, i.e., $\ell = 2^N$ and $\log d \in \mathbb{N}^+$. This setting aligns with the scaling of quantum computing, making it easier to understand. For other cases, padding techniques can be applied to expand the matrix and vector dimensions to conform to this requirement.

Since the runtime speedups depend heavily on the capabilities of the available input oracles, it is essential to specify the input oracles used in the quantum Transformer before detailing the algorithms. For the classical Transformers, memory access to the inputs such as the sentence and the query, key, and value matrices is assumed. In the quantum version, access to several matrices via *block encoding techniques* introduced in Sect. 2.4 is assumed.

Assumption 5.1 (Input Oracles) *Following the explicit form of the single-head and single-layer Transformer in Eq. (5.5), there are five parameterized weight matrices, i.e., W_q, W_k, $W_v \in \mathbb{R}^{d \times d}$ in the attention block, as well as $M_1 \in \mathbb{R}^{d' \times d}$ and $M_2 \in \mathbb{R}^{d \times d'}$ in FFN. Note that here, M_1 and M_2 are actually the transpose of parameterized matrices in the classical transformer. The quantum Transformer assumes access to these five parameterized weight matrices, as well as the input sequence $S \in \mathbb{R}^{\ell \times d}$ via block encoding.*

Mathematically, given any $A \in \{W_q, W_k, W_v, M_1, M_2, S\}$ corresponding to an N-qubit operator, $\alpha, \varepsilon \geq 0$ and $a \in \mathbb{N}$, there exists a $(a + N)$-qubit unitary U_A referring to the (α, a, ε)-block encoding of A with

$$\| A - \alpha(\langle 0|^{\otimes a} \otimes \mathbb{I}_{2^N}) U_A (|0\rangle^{\otimes a} \otimes \mathbb{I}_{2^N}) \| \leq \varepsilon, \tag{5.7}$$

where $\| \cdot \|$ represents the spectral norm.

Under this assumption, the quantum Transformer can access U_S, U_{W_q}, U_{W_k}, and U_{W_v} corresponding to the (α_s, a_s)-encoding of S and (α_w, a_w)-encodings of W_q, W_k, and W_v, respectively. Moreover, the quantum Transformer can access (α_m, a_m)-encodings U_{M_1} and U_{M_2} corresponding two weight matrices $M_1 \in \mathbb{R}^{d' \times d}$ and $M_2 \in \mathbb{R}^{d \times d'}$ in FFN.

> **Remark**
>
> For simplicity and clarity, in the following, the *perfect* block encoding of input matrices *without errors* is considered, i.e., $\varepsilon = 0$. As such, the error term of the block encoding will not be explicitly written, and (α, a) will be used instead of $(\alpha, a, 0)$. The output of the quantum Transformer is a quantum state corresponding to the probability distribution of the next token.

The complete single-layer structure is described in Fig. 5.2. A quantum Transformer consists of a self-attention and a feed-forward network sub-layer, incorporating residual connections with layer normalization. The inputs of the quantum Transformer are block encodings of matrices for the input sequence and pretrained weights, from which the relevant matrices for the transformer are constructed (query Q, key K, and value V). Each of the components accepts the block encoding from the prior component as the input and prepares a block encoding of the target matrix using quantum linear algebra as the output.

Under the above assumptions about access to the read-in protocols, the following theorem indicates how to implement a single-head and single-block Transformer architecture in Eq. (5.5) on the quantum computer.

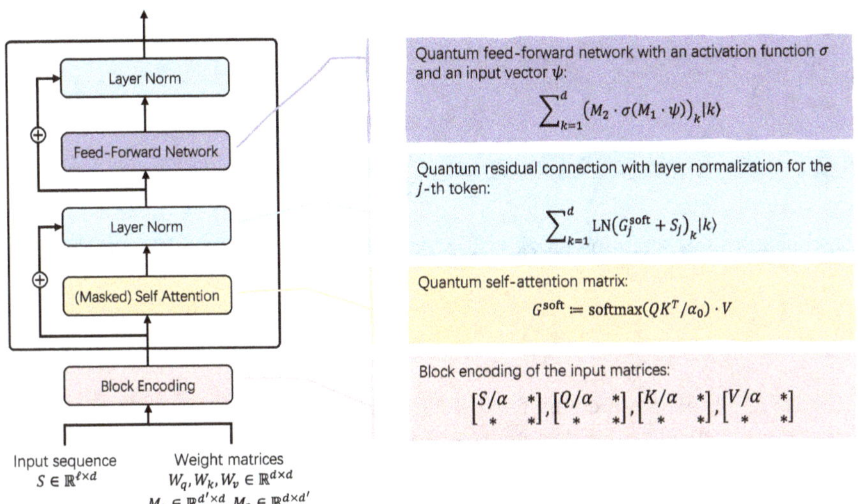

Fig. 5.2 Overview of the single-layer decoder-only quantum transformer

Theorem 5.1 (Quantum Transformer, Informal) *For a single-head and single-block Transformer depicted in Fig. 5.2, suppose its embedding dimension is d and its input sequence S has the length $\ell = 2^N$. Under Assumption 5.1 about the input oracles, for the index $j \in [\ell]$, one can construct an ϵ-accurate quantum circuit for the quantum state proportional to*

$$\sum_{k=1}^{d} \mathrm{Transformer}(S, j)_k |k\rangle, \tag{5.8}$$

by using $\tilde{O}\!\left(dN^2 \alpha_s \alpha_w \log^2(1/\epsilon)\right)$ times of the input block encodings.

This theorem is demonstrated by explicitly designing the quantum circuit for each computation block of the Transformer architecture in a coherent way, i.e., without intermediate measurement. In addition, a subsequent transformation of the amplitude-encoded state, followed by measurement in the computational basis, yields the index of the next predicted token based on the probabilities modeled by the Transformer architecture.

Roadmap In the remainder of this section, we detail the implementation of quantum Transformers, proceeding from the bottom to the top as illustrated in Fig. 5.2. Specifically, the quantization of the attention block Attention(S, j) is first demonstrated in Sect. 5.2.1. Next, the quantization of residual connections and layer normalization operations (i.e., the operation LN(Attention(S, j)) in Eq. (5.5)) is presented in Sect. 5.2.2. Last, the quantization of the fully connected neural network is exhibited to complete the computation LN(FFN(LN(Attention(S, j)))))

in Sect. 5.2.3. This end-to-end approach ensures that the generated quantum state corresponds to the one described in Eq. (5.8).

5.2.1 Quantum Self-Attention

The quantum self-attention block is now described, aiming to complete the computation:

$$\text{Attention}(Q, K, V) = \text{softmax}\left(QK^\top/\alpha_0\right)V =: G^{\text{soft}}$$

in Eq. (5.1) on quantum computers. More specifically, under Assumption 5.1, the quantum self-attention block outputs a block encoding of a matrix G whose j-th row is the same as the output of the classical attention block, as described in the following theorem.

Theorem 5.2 (Quantum Self-Attention) *Let $\alpha_0 = \alpha_s^2\alpha_w^2$. For the index $j \in [\ell]$, one can construct a block encoding of a matrix G such that $G_{j\star} = G_j^{\text{soft}} :=$ $(\text{softmax}\left(QK^\top/\alpha_0\right)V)_{j\star}$.*

> **Remark**
> For quantum self-attention, a slight change is made by setting the scaling factor $\alpha_0 = \alpha_s^2\alpha_w^2$ for the following reasons. The first is that the usual setting $\alpha_0 = 1/\sqrt{d}$ is chosen somewhat heuristically, and there are already some classical works considering different scaling coefficients which may even achieve better performance [13, 14]. The second, which is more important, is that the quantum input assumption using the block encoding format naturally contains the normalization factor α which plays a similar role to the scaling factor. Therefore, for the quantum case in the context of this work, it suffices to use α directly.

The implementation of the quantum self-attention block can be decomposed into three steps:

1. Construction of the block encoding of the matrix $QK^\top$ and the matrix V given access to U_S, U_{W_q}, U_{W_k}, and U_{W_v}.
2. Implementation of the quantum algorithm to compute the softmax function $\text{softmax}\left(QK^\top/\alpha_0\right)$ given access to U_{KQ}.
3. Multiply with V via the product of block encodings.

In what follows, the implementation of each step is iteratively detailed.

Step I The construction of the block encoding of matrix $QK^\top$ and V builds upon the employment of Fact 2.3. That is, given access to (α, a)-encoding U_A of matrix A and (β, b)-encoding U_B of matrix B, an $(\alpha\beta, a+b)$-encoding can be constructed for the matrix AB. This result leads to the efficient construction of the block encodings of $QK^\top$ and V asz summarized below.

- For the matrix $V = W_v S$, it is straightforward to set $A = W_v$ and $B = S$ to construct the (α_v, a_v)-encoding U_V, where $\alpha_v = \alpha_s \alpha_w$ and $a_v = a_s + a_w$.
- For the matrix $QK^\top$, we first use Fact 2.3 to construct the (α_q, a_q)-encoding U_Q and (α_k, a_k)-encoding U_K with $Q = W_q S$ and $K = W_k S$, respectively. Then, we use Fact 2.3 again to construct the (α_0, a_0)-encoding $U_{QK^\top}$ of $QK^\top$, where $\alpha_0 = \alpha_s^2 \alpha_w^2$ and $a_0 = 2a_s + 2a_w$. Note that for a real matrix M and its block encoding unitary U_M, $U_M^\dagger$ is the block encoding of $M^\top$.

Step II Once the unitary $U_{QK^\top}$ is prepared, the quantum algorithm corresponding to the softmax function, i.e., $\mathrm{softmax}(QK^\top/\alpha_0)$, is implemented. Note that the softmax function relies on the exponential function, which is generally resource-intensive to implement on quantum computers. To circumvent this bottleneck, the quantum Transformer uses polynomial functions to approximate the softmax function, as supported by the following fact.

Fact 5.1 *For $x \in [-1, 1]$, the function $f(x) := e^x$ can be approximated with error bound ϵ with an $O(\log(1/\epsilon))$-degree polynomial function.*

The insight provided by Fact 5.1 is to use polynomial functions to approximate the softmax function, i.e., $\exp \circ (QK^\top/\alpha_0)$ is first approximated using polynomial functions and then multiplied with different coefficients (normalization) for each row.

> **Remark**
> The notation $\exp \circ (A)$ indicates that the exponential operation is applied element-wise to each entry of the matrix A, rather than representing a matrix exponential.
> Moreover, the element-wise functions mean that functions are implemented on each matrix element.

In this context, the challenge of implementing a quantized softmax function reduces to the implementation of a quantized polynomial function. The key technique for achieving this lies in applying polynomial functions to each element of block-encoded matrices, as detailed in the following lemma.

Lemma 5.1 (Element-Wise Polynomial Function of Block-Encoded Matrix)
Let $N, k \in \mathbb{N}$. Given access to an (α, a)-encoding of a matrix $A \in \mathbb{C}^{2^N \times 2^N}$ and

an r-degree polynomial function $f_r(x) = \sum_{j=1}^{r} c_j x^j$, $c_j \in \mathbb{C}$ for $j \in [r]$, one can construct a (C, b)-encoding of $f_r \circ (A/\alpha)$ by using $O(r^2)$ times the input unitary, where $C := \sum_{j=1}^{r} |c_j|$, $b := ra + (r-1)N + \lceil \log(r+1) \rceil$.

Moreover, for a polynomial function $g_r(x) = \sum_{j=0}^{r} c_j x^j$ with constant term c_0, one can construct a (C', b)-encoding of $g_r \circ (A/\alpha)$, where $C' = rc_0 + C$.

Proof of Lemma 5.1 To achieve this implementation, two state-preparation unitaries P_L and P_R are constructed, which act on $\lceil \log(r+1) \rceil$ qubits such that

$$P_L : \; |0^{\lceil \log(r+1) \rceil}\rangle \to \frac{1}{\sqrt{C}} \sum_{j=1}^{r} \sqrt{|c_j|}|j\rangle, \tag{5.9}$$

$$P_R : \; |0^{\lceil \log(r+1) \rceil}\rangle \to \frac{1}{\sqrt{C}} \sum_{j=1}^{r} \sqrt{|c_j|}e^{i\theta_j}|j\rangle, \tag{5.10}$$

where $C = \sum_{j=1}^{r} |c_j|$ and $|c_j|e^{i\theta_j} = c_j$. These two unitaries encode the polynomial coefficients $\{c_j\}$ into the quantum circuit, which is needed for block encoding via the linear combination of unitaries indicated in Fact 2.2. Note that the construction of P_L and P_R is efficient for small r, as the corresponding circuit is $O(r)$-depth with only elementary quantum gates [15, 16].

For $j \in [r]$, let U_{A^j} be the $(1, ja + (j-1)N)$-encoding of

$$(A/\alpha)^{\circ j} := \underbrace{(A/\alpha) \circ (A/\alpha) \circ \cdots \circ (A/\alpha)}_{j-1 \text{ times of Hadamard product}},$$

which is constructed by iteratively applying Lemma 2.1. Now, the construction of the unitary $W = \sum_{j=1}^{r} |j\rangle\langle j| \otimes U_{A^j} + (I_{2\log r} - \sum_{j=1}^{r} |j\rangle\langle j|) \otimes I_{ra+rn}$ is described. Instead of preparing block encodings of $A^{\circ j}$ for all $j \in [r]$, it suffices to prepare block encodings of $A^{\circ 2^j}$ for $j \in \lfloor \log r \rfloor$. As an example, $A^{\circ 7} = A^{\circ 4} \cdot A^{\circ 2} \cdot A^{\circ 1}$. Combining these together, we need to use $O\left(\sum_{j=1}^{\lfloor \log r \rfloor} 2^j\right) = O(r)$ times of U_A to construct $(ra + (r-1)N + 2\log r)$-qubit unitary W. By Fact 2.2, a $(C, ra + (r-1)N + 2\log r)$-encoding of $f_r \circ (A/\alpha)$ can be implemented.

To implement element-wise functions including constant terms, access to the block encoding of a matrix whose elements are all 1 is also needed. Notice that this matrix can be written as the linear combination of the identity matrix and the reflection operator, i.e.,

$$\sum_{k,k' \in [2^N]} |k\rangle\langle k'| = \frac{2^N}{2}\left(\mathbb{I}_{2^N} - \left(\mathbb{I}_{2^N} - \frac{2}{2^N}\sum_{k,k' \in [2^N]} |k\rangle\langle k'|\right)\right) \tag{5.11}$$

$$= \frac{2^N}{2}\left(\mathbb{I}_{2^N} - H^{\otimes N}\left(\mathbb{I}_{2^N} - 2|0^N\rangle\langle 0^N|\right)H^{\otimes N}\right), \tag{5.12}$$

where H is the Hadamard gate. Define $U_{\text{ref}} = |0\rangle\langle 0| \otimes \mathbb{I}_{2^N} + |1\rangle\langle 1| \otimes \left(H^{\otimes N}(\mathbb{I}_{2^N} - 2|0^N\rangle\langle 0^N|)H^{\otimes N}\right)$. By direct computation, one can show that $U_0 = (XH \otimes \mathbb{I}_{2^N})U_{\text{ref}}(H \otimes \mathbb{I}_{2^N})$ is an $(2^N, 1)$-encoding of $\sum_{k,k'} |k\rangle\langle k'|$. One can achieve the element-wise function by following the same steps as above and taking linear combinations among $U_0, \ldots, U_{A^r}$. A point to notice is that only $(2^N, 1)$-encoding of the matrix whose elements are all 1 can be constructed since the spectral norm of this matrix is 2^N. Therefore, $2^N c_0$ s encoded into the state instead of c_0 to amplify the constant term. $\qquad\square$

Supported by the above lemma, Step II (i.e., the quantum softmax for self-attention) can be completed, as shown in the following theorem.

Theorem 5.3 (Quantum Softmax for Self-Attention, Informal) *Given an (α, a)-encoding U_A of a matrix $A \in \mathbb{R}^{\ell \times \ell}$, a positive integer d, and an index $j \in [\ell]$, one can prepare a state encoding of*

$$|A_j\rangle := \sum_{k=1}^{\ell} \sqrt{\text{softmax}\,(A/\alpha)_{jk}}|k\rangle = \frac{1}{\sqrt{Z_j}} \sum_{k=1}^{\ell} \exp \circ \left(\frac{A}{2\alpha}\right)_{jk}|k\rangle,$$

where $Z_j = \sum_{k=1}^{\ell} \exp \circ (A/\alpha)_{jk}$.

Proof Sketch of Theorem 5.3 First, the block encoding of $\exp \circ(\frac{A}{2\alpha})$ is constructed. Note that Taylor expansion of $\exp(x)$ contains a constant term 1. This can be achieved with Lemma 5.1 and Fact 5.1. Here, since we are only focusing on the j-th row, instead of taking linear combination with the matrix whose elements are all 1, we take sum with the matrix whose j-th row elements are all 1 and else are 0. This enables us to have a better dependency on ℓ, i.e., from ℓ to $\sqrt{\ell}$. For index $j \in [\ell]$, let $U_j : |0\rangle \to |j\rangle$. One can achieve this by changing Eq. (5.12) to the following:

$$\sum_k |j\rangle\langle k| = \frac{\sqrt{\ell}}{2}\left(U_j H^{\otimes N} - U_j \left(\mathbb{I}_{2^N} - 2|0^N\rangle\langle 0^N|\right) H^{\otimes N}\right). \tag{5.13}$$

Following the same steps in Lemma 5.1, one can achieve the construction. There are two error terms in this step. Denote $U_{f\circ(A)}$ as the constructed block encoding unitary. By Lemma 5.1 and some additional calculation, one can show that $U_{f\circ(A)}$ is a block encoding of $\exp \circ(\frac{A}{2\alpha})$. Note that $\exp \circ(\frac{A}{2\alpha})_{jk} = \exp \circ(\frac{A}{2\alpha})_{kj}^{\top}$. With unitary $U_{f\circ(A)}^{\dagger}(I \otimes U_j)$ and amplitude amplification, one can prepare a state encoding of the target state

$$|A_j\rangle := \frac{1}{\sqrt{Z_j}} \sum_{k=1}^{\ell} \exp \circ \left(\frac{A}{2\alpha}\right)_{jk}|k\rangle, \tag{5.14}$$

where $Z_j = \sum_{k=1}^{\ell} \exp \circ (A/\alpha)_{jk}$ is the normalization factor of softmax function for the j-th row. $\square$

Step III Finally, the matrix multiplication with V is implemented. This can be easily achieved by using Fact 2.3, with $U^{\dagger}_{f(QK^{\top})}$ and U_V. Consequently, an encoded quantum state is obtained, analogous to

$$\sum_{k}^{\ell} \left(\mathrm{softmax}\left(QK^{\top}/\alpha_0\right)V\right)_{jk}|k\rangle. \tag{5.15}$$

Combining the results of Steps I, II, and III, the proof of Theorem 5.2 can be presented below.

Proof of Theorem 5.2 In the first step, the block encoding of matrix $QK^{\top}$ and V is constructed. Note that for a real matrix M and its block encoding unitary U_M, $U^{\dagger}_M$ is the block encoding of $M^{\top}$. By Fact 2.3, one can construct an (α_0, a_0)-encoding $U_{QK^{\top}}$ of $QK^{\top}$, where $\alpha_0 := \alpha_s^2 \alpha_w^2$ and $a_0 = 2a_s + 2a_w$. One can also construct an (α_v, a_v)-encoding U_V of V, where $\alpha_v = \alpha_s \alpha_w$ and $a_v = a_s + a_w$.

By Theorem 5.3, using $U_{QK^{\top}}$, one can prepare a state encoding of the state:

$$\sum_{k=1}^{\ell} \sqrt{\mathrm{softmax}\left(QK^{\top}/\alpha_0\right)_{jk}}|k\rangle,$$

where $Z_j = \sum_{k=1}^{\ell} \exp \circ (QK^{\top}/\alpha_0)_{jk}$. Remember that state encoding is also a block encoding. By Lemma 2.1, one can construct a block encoding of a matrix whose j-th column is

$$\left[\mathrm{softmax}\left(QK^{\top}/\alpha_0\right)_{j1}, \ldots, \mathrm{softmax}\left(QK^{\top}/\alpha_0\right)_{j\ell}\right]$$

ignoring other columns. Let this block encoding unitary be $U_{f(QK^{\top})}$.

Last, by exploiting Fact 2.3 again, with $U^{\dagger}_{f(QK^{\top})}$ and U_V, we obtain an encoded quantum state analogous to

$$\sum_{k}^{\ell} \left(\mathrm{softmax}\left(QK^{\top}/\alpha_0\right)V\right)_{jk}|k\rangle.$$

$\square$

5.2.1.1 Extension to Implement Quantum Masked Self-Attention

This section considers the implementation of the *masked* self-attention, which is essential for the decoder-only structure. This can be achieved by slightly changing some steps as introduced in previous theorems.

Corollary 5.1 (Quantum Masked Self-Attention) *For the index $j \in [\ell]$, one can construct a block encoding of a matrix G^{mask} such that $G_{j\star}^{\mathrm{mask}} = \left(\mathrm{softmax}\left(\frac{QK^{\top}}{\alpha_0} + M \right) V \right)_{j\star}$, where M is the masked matrix as Eq. (5.2), $Z_j = \sum_{k=1}^{\ell} \exp \circ \left(\frac{QK^{\top}}{\alpha_0} + M \right)_{jk}$.*

Proof of Corollary 5.1 To achieve masked self-attention, the steps mentioned in Theorem 5.3 are slightly modified.

First, to approximate the exponential function, the approach of using a matrix where all elements in the j-th row are set to 1 while other rows remain 0 is extended. Instead, this approach is refined by considering only the first $2^{\lceil \log(j+1) \rceil}$ elements in the j-th row to be 1. Note that this matrix can be achieved similarly to the original one. The encoding factor of this matrix is $2^{\lceil \log(j+1) \rceil / 2}$.

Second, after approximating the function, for index $j \in [\ell]$, the block encoding is multiplied with a projector $\sum_{k,k \leq j} |k\rangle \langle k|$ to mask the elements. Though the projector $\sum_{k \in S} |k\rangle \langle k|$ for $S \subseteq [\ell]$ is not unitary in general, one can construct a block encoding of the projector as it can be written by the linear combination of two unitaries:

$$\sum_{k \in S} |k\rangle \langle k| = \frac{1}{2}\mathbb{I} + \frac{1}{2}\left(2\sum_{k \in S} |k\rangle \langle k| - \mathbb{I} \right). \tag{5.16}$$

Define $U_{\mathrm{proj}} := |0\rangle \langle 0| \otimes \mathbb{I} + |1\rangle \langle 1| \otimes (2 \sum_{k \in S} |k\rangle \langle k| - \mathbb{I})$. One can easily verify that $(H \otimes \mathbb{I}) U_{\mathrm{proj}} (H \otimes \mathbb{I})$ is a $(1, 1, 0)$-encoding of $\sum_{k \in S} |k\rangle \langle k|$, where H is the Hadamard gate. The following steps follow the same with Theorems 5.2 and 5.3. $\square$

One may further achieve the multihead self-attention case by using the linear combination of unitaries.

5.2.2 Quantum Residual Connection and Layer Normalization

This subsection discusses the implementation of the residual connection with layer normalization on a quantum computer. The implementation of the layer norm block, as shown in Fig. 5.2, continues based on the result in Theorem 5.2.

Theorem 5.4 (Quantum Residual Connection with Layer Normalization)
Given access to the block encoding of the matrix G in Theorem 5.2, one can construct a quantum-encoded state:

$$\sum_{k=1}^{d} \mathrm{LN}\big(G_j^{\mathrm{soft}}, S_j\big)_k |k\rangle = \frac{1}{\varsigma} \sum_{k=1}^{d} \big(G_{jk}^{\mathrm{soft}} + S_{jk} - \bar{s}_j\big)|k\rangle,$$

where $\bar{s}_j := \frac{1}{d} \sum_{k=1}^{d} \big(G_{jk}^{\mathrm{soft}} + S_{jk}\big)$ *and* $\varsigma := \sqrt{\sum_{k=1}^{d} \big(G_{jk}^{\mathrm{soft}} + S_{jk} - \bar{s}_j\big)^2}$.

Proof of Sketch of Theorem 5.4 As shown in Theorem 5.2, a block encoding of a matrix G whose j-th row is the same row as that of G^{soft} can be constructed. By Assumption 5.1, U_s, an (α_s, a_s)-encoding of S, is provided. By Lemma 2.4 with the state preparation pair (P, P) such that

$$P|0\rangle = \frac{1}{\sqrt{\alpha_g + \alpha_s}}\big(\sqrt{\alpha_g}|0\rangle + \sqrt{\alpha_s}|1\rangle\big), \tag{5.17}$$

one can construct a quantum circuit U_{res} which is an $(\alpha_g + \alpha_s, a_g + 1)$-encoding of an $\ell \times d$ matrix whose j-th row is the same as that of $G^{\mathrm{soft}} + S$.

Now, the creation of a block encoding of a diagonal matrix $\bar{s}_j \cdot \mathbb{I}$, where $\bar{s}_j := \frac{1}{d} \sum_{k=1}^{d} \big(G_{jk}^{\mathrm{soft}} + S_{jk}\big)$, is considered. Define a unitary as $H^{\log d} := H^{\otimes \log d}$. Note that $H^{\log d}$ is a $(1, 0, 0)$-encoding of itself, and the first column of $H^{\log d}$ is $\frac{1}{\sqrt{d}}(1, \ldots, 1)^\top$. By Fact 2.3, one can multiply $G^{\mathrm{soft}} + S$ with $H^{\log d}$ to construct a block encoding of an $\ell \times d$ matrix, whose $(j, 1)$-element is $\sqrt{d}\bar{s}_j$. One can further move this element to $(1, 1)$ by switching the first row with the j-th row. By tensor product with the identity $\mathbb{I}$ of $\log d$ qubits, one can construct a block encoding of $\sqrt{d}\bar{s}_i \cdot \mathbb{I}$.

With $U_j : |0\rangle \to |j\rangle$, one can prepare the state:

$$U_{\mathrm{res}}^\dagger(\mathbb{I} \otimes U_j)|0\rangle|0\rangle = \frac{1}{\alpha_g + \alpha_s}|0\rangle \sum_{k=1}^{d} \psi_k |k\rangle + \sqrt{1 - \frac{\sum_k \psi_k^2}{(\alpha_g + \alpha_s)^2}}|1\rangle|\mathrm{bad}\rangle. \tag{5.18}$$

By the diagonal block encoding of amplitudes mentioned as Fact 2.6, this can be converted to a block encoding of the diagonal matrix $\mathrm{diag}(G_{j1} + S_{j1}, \ldots, G_{jd} + S_{jd})$.

By taking the linear combination as Fact 2.4 with state preparation pair (P_1, P_2), where

$$P_1|0\rangle = \frac{1}{\sqrt{1 + 1/\sqrt{d}}}\left(|0\rangle + \frac{1}{\sqrt{d}}|1\rangle\right) \tag{5.19}$$

and

$$P_2|0\rangle = \frac{1}{\sqrt{1+1/\sqrt{d}}}\left(|0\rangle - \frac{1}{\sqrt{d}}|1\rangle\right), \qquad (5.20)$$

one can construct a block encoding of $\mathrm{diag}\left(G_{j1} + S_{j1} - \bar{s}_j, \ldots, G_{jd} + S_{jd} - \bar{s}_j\right)$, and we call it U_{LN}. Then, the unitary $U_{\mathrm{LN}}\left(\mathbb{I} \otimes H^{\log d}\right)$ is an state encoding of the state

$$\frac{1}{\varsigma}\sum_{k=1}^{d}\left(G_{jk}^{\mathrm{soft}} + S_{jk} - \bar{s}_j\right)|k\rangle,$$

where $\varsigma := \sqrt{\sum_{k=1}^{d}\left(G_{jk}^{\mathrm{soft}} + S_{jk} - \bar{s}_j\right)^2}$. $\qquad\qquad\qquad\square$

5.2.3 Quantum Feed-Forward Neural Network

Attention is now turned to the third main building block of the transformer architecture, the feed-forward neural network. This block often is a relatively shallow neural network with linear transformations and ReLU activation functions [1]. More recently, activation functions such as the GELU have become popular, being continuously differentiable. We highlight that they are ideal for quantum Transformers, since the QSVT framework requires functions that are well approximated by polynomial functions. Functions like $\mathrm{ReLU}(x) = \max(0, x)$ cannot be efficiently approximated. The GELU is constructed from the error function, which is efficiently approximated as follows.

Fact 5.2 (Polynomial Approximation of Error Function [17]) *Let $\epsilon > 0$. For every $k > 0$, the error function $\mathrm{erf}(kx) := \frac{2}{\sqrt{\pi}}\int_0^{kx} e^{-t^2}\, dt$ can be approximated with error up to ϵ by a polynomial function with degree $O\left(k\log\left(\frac{1}{\epsilon}\right)\right)$.*

This lemma implies the following efficient approximation of the GELU function with polynomials.

Corollary 5.2 (Polynomial Approximation of GELU Function) *Let $\epsilon > 0$ and $\lambda \in O(1)$. For every $k > 0$ and $x \in [-\lambda, \lambda]$, the GELU function $\mathrm{GELU}(kx) := kx \cdot \frac{1}{2}\left(1 + \mathrm{erf}\left(\frac{kx}{\sqrt{2}}\right)\right)$ can be approximated with error up to ϵ by a polynomial function with degree $O\left(k\log\left(\frac{k\lambda}{\epsilon}\right)\right)$.*

Proof of Corollary 5.2 It suffices to approximate the error function with precision $\frac{\epsilon}{k\lambda}$ by Fact 5.2. $\qquad\qquad\qquad\qquad\qquad\qquad\qquad\qquad\square$

The following theorem considers the implementation of the two-layer feed-forward network on quantum computers. As mentioned, the GELU function is widely used in transformer-based models and is explicitly considered as the activation function in the theorem. Cases for other activation functions like sigmoid follow the same analysis. An example is the $\tanh(x)$ function, which can be well approximated by a polynomial for $x \in [-\pi/2, \pi/2]$ [18].

Theorem 5.5 (Two-Layer Feed-Forward Network with GELU Function, Informal) *Assume we have access to (α, a)-state encoding of an N-qubit state $|\psi\rangle = \sum_{k=1}^{2^N} \psi_k |k\rangle$, where $\{\psi_k\}$ are real and $\|\psi\|_2 = 1$. Further, assume access to (α_m, a_m)-encodings U_{M_1} and U_{M_2} of weight matrices $M_1 \in \mathbb{R}^{d' \times d}$ and $M_2 \in \mathbb{R}^{d \times d'}$. Let the activation function be $\mathrm{GELU}(x) := x \cdot \frac{1}{2}\left(1 + \mathrm{erf}\left(\frac{x}{\sqrt{2}}\right)\right)$. One can prepare a state encoding of the state:*

$$|\phi\rangle = \frac{1}{C} \sum_{k=1}^{d} \left(M_2 \cdot \mathrm{GELU}(M_1 \cdot \psi) \right)_k |k\rangle, \tag{5.21}$$

where C is the normalization factor.

Proof of sketch of Theorem 5.5 The proof proceeds as follows. Recall that

$$(\mathbb{I}_{2^a} \otimes U_{M_1})(\mathbb{I}_{2^{a_m}} \otimes U_\psi)|0^{a+a_m+N}\rangle = \frac{1}{\alpha\alpha_m}|0^{a+a_m}\rangle M_1|\psi\rangle + |\widetilde{\perp}\rangle, \tag{5.22}$$

where $|\widetilde{\perp}\rangle$ is an unnormalized orthogonal state. For the case $d' \geq \ell$, this can be achieved by padding ancilla qubits to the initial state. By Fact 2.6, one can construct a block encoding of the diagonal matrix $\mathrm{diag}((M_1\psi)_1, \ldots, (M_1\psi)_{d'})$. Note that the GELU function does not have a constant term and is suitable to use the importance-weighted amplitude transformation as in [19]. Instead of directly implementing the GELU function, the function $f(x) = \frac{1}{2}(1 + \mathrm{erf}(\frac{x}{\sqrt{2}}))$ is first implemented. Note that the value of $|\mathrm{erf}(x)|$ is upper bounded by 1. By Fact 2.6 with function $\frac{1}{4}(1 + \mathrm{erf}(\alpha\alpha_m\frac{x}{\sqrt{2}}))$, one can construct a block encoding of matrix $\mathrm{diag}\left(f(M_1\psi)_1, \ldots, f(M_1\psi)_{d'}\right)$.

Let the previously constructed block encoding unitary be $U_{f(x)}$. We have

$$U_{f(x)}(\mathbb{I} \otimes U_{M_1})(\mathbb{I} \otimes U_\psi)|0\rangle|0\rangle = \frac{1}{2\alpha\alpha_m}|0\rangle \sum_k \mathrm{GELU}(M_1\psi)_k |k\rangle + |\widetilde{\perp'}\rangle,$$

$$\tag{5.23}$$

where $|\widetilde{\perp}'\rangle$ is an unnormalized orthogonal state. Finally, by implementing the block encoding unitary U_{M_2}, the result is

$$(\mathbb{I} \otimes U_{M_2})(\mathbb{I} \otimes U_{f(x)})(\mathbb{I} \otimes U_{M_1})(\mathbb{I} \otimes U_\psi)|0\rangle|0\rangle$$

$$= \frac{C}{2\alpha\alpha_m^2}|0\rangle \sum_{k=1}^{d} \Big(M_2 \cdot \mathrm{GELU}(M_1 \cdot \psi) \Big)_k |k\rangle + |\widetilde{\perp}''\rangle, \tag{5.24}$$

where C is the normalization factor and $|\widetilde{\perp}''\rangle$ is an unnormalized orthogonal state.

$\square$

Remark
The quantum feed-forward network discussed in this subsection is a quantum implementation of the classical feed-forward network under the input assumption of block encoding, which is essentially different from the quantum analog of neural networks introduced in Chap. 4.

5.3 Runtime Analysis with Quadratic Speedups

This section provides a combined analytical and numerical analysis to explore the potential of a quantum speedup in time complexity.

5.3.1 Overview

With the quantum implementation of self-attention, residual connection, layer normalization, and feed-forward networks, the quantum transformer can be constructed by combining these building blocks as in Theorem 5.1.

This final complexity is obtained on the basis of the following considerations: the single-head and single-block transformer architecture includes one self-attention, one feed-forward network, and two residual connections with layer normalization, as shown in Fig. 5.2.

Starting from the input assumption as Assumption 5.1, for the index $j \in [\ell]$, the block encoding of the self-attention matrix is first constructed, as described in Sect. 5.2.1. This output can be directly the input of the quantum residual connection and layer normalization, as Sect. 5.2.2, which output is a state encoding. Remind the definition of state encoding as Definition 2.10. The state encoding can directly be used as the input of the feed-forward network, as Sect. 5.2.3. Finally, we put the output of the feed-forward network, which is a state encoding, into the residual connection block. This is possible by noticing that state encoding is a specific kind of block encoding. Multiplying the query complexity of these computational blocks, one can achieve final result. The detailed analysis of runtime is referred to Guo et al. [20]

As Theorem 5.1 shows, the quantum transformer uses $\tilde{O}(dN^2\alpha_s\alpha_w)$ times the input block encodings, where α_s and α_w are encoding factors. By analyzing naive matrix multiplication, the runtime of classical single-head and single-block Transformer during the inference stage is $O(\ell d + d^2)$, where d is the embedding dimension and $\ell = 2^N$ is the input sequence length. From the comparison, it can be seen that α_s and α_w are the dominant factors that affect the potential quantum speedup. The properties of these two factors will be explored via numerical studies.

5.3.2   Empirical Studies of Potential Quantum Speedups

The encoding factors α_s and α_w appear in the block encodings of matrices S and W_q, W_k, W_v. Recall that the encoding factor α is lower bounded by the spectral norm of a block-encoded matrix A, i.e., $\alpha \geq \|A\|$. Given access to quantum random access memory (QRAM) and a quantum data structure [21, 22], there are well-known implementations that enable the construction of a block encoding for an arbitrary matrix A where the encoding factor is upper bounded by the Frobenius norm $\|A\|_F$. Based on these considerations, these two norms of the input matrices of several open-source large language models are numerically studied[1] to provide upper and lower bound of α_s and α_w.

The input sequence matrix S, which introduces the dependency on ℓ, is first investigated. Input data in real-world applications sampled from the widely used Massive Multitask Language Understanding (MMLU) dataset [23] are considered, covering 57 subjects across science, technology, engineering, mathematics, the humanities, the social sciences, and more. The scaling of the spectral norm and Frobenius norm of S on the MMLU dataset is demonstrated in Fig. 5.3. It can be found that the matrix norms of the input matrix of all LLMs scale at most as $O(\sqrt{\ell})$.

As additional interest, this analysis of the matrix norm provides new insights for the classical tokenization and embedding design. It can be observed that comparatively more advanced LLM models like *Llama2-7b* and *Mistral-7b* have large variances of matrix norms. This phenomenon is arguably the consequence of the training in those models; the embeddings that frequently appear in the real-world dataset are actively updated at the pretraining stage and, therefore, are more broadly distributed.

---

[1] Parameters are obtained from the Hugging Face website, which is an open-source platform for machine learning models.

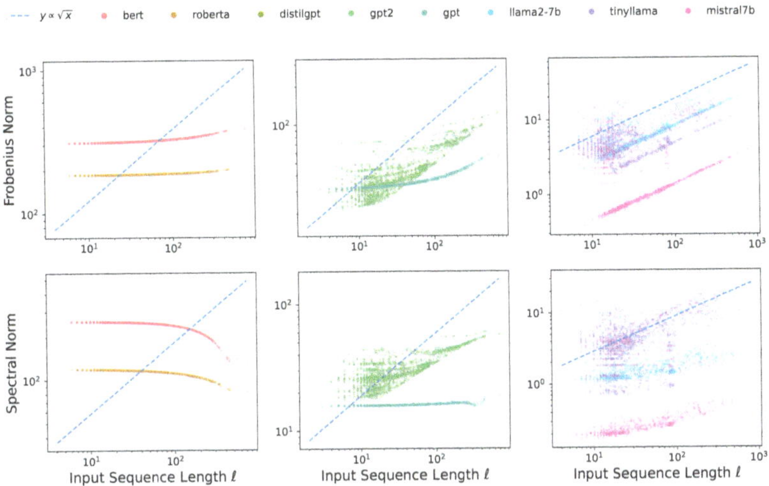

Fig. 5.3 Scaling of the spectral norm $\|S\|$ and the Frobenius norm $\|S\|_F$ with ℓ for each model, displayed on logarithmic scales for both axes

The spectral and Frobenius norms of weight matrices (W_q, W_k, W_v) for the large language models are then computed. The result can be seen in Fig. 5.4. Many of the LLMs below a dimension d of 10^3 that we have checked have substantially different norms. It is observed that for larger models such as *Llama2-7b* and *Mistral-7b*, which are close to the current state-of-the-art open-source models, the norms do not change dramatically. Therefore, it is reasonable to assume that the spectral norm and the Frobenius norm of the weight matrices are at most $O(\sqrt{d})$ for advanced LLMs.

Given these numerical experiments, it is reasonable to assume that $\alpha_s = O(\sqrt{\ell})$ and $\alpha_w = O(\sqrt{d})$, and we obtain a query complexity of the quantum transformer in $\widetilde{O}(d^{\frac{3}{2}}\sqrt{\ell})$. We continue with a discussion of the possible time complexity. With the QRAM assumption, the input block encodings can be implemented in a polynomially logarithmic time of ℓ. Even without a QRAM assumption, there can be cases when the input sequence is generated efficiently, for example, when the sequence is generated from a differential equation, see additional discussions in the supplementary material. In these cases, we demonstrate that a quadratic speedup of the runtime of a single-head and single-block Transformer can be expected.

5.4 Code Demonstration

This section explains how self-attention works with a simple concrete example. Consider a short sequence of three words: "the cat sleeps."

First, we convert each word into an embedding vector. For this toy example, very small four-dimensional embeddings are used:

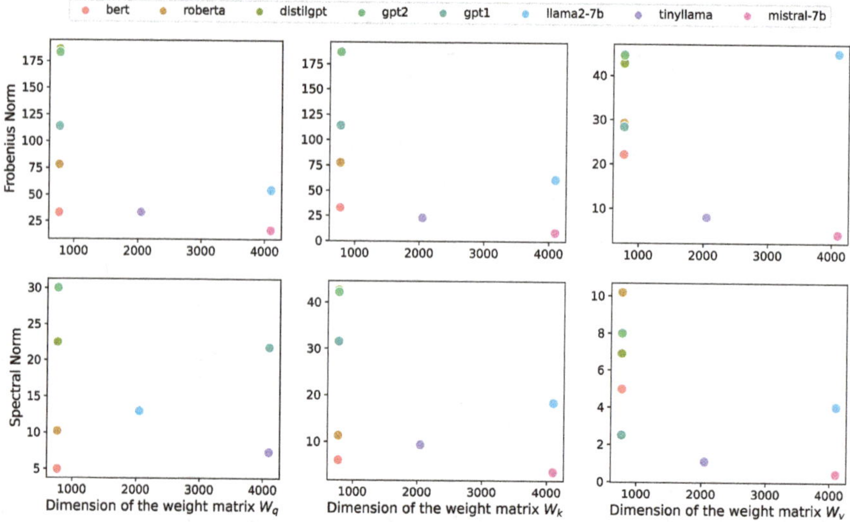

Fig. 5.4 Norms of weight matrices across open-source LLMs

```
The     = [1, 0, 1, 0]
cat     = [0, 1, 1, 1]
sleeps  = [1, 1, 0, 1]
```

In self-attention, each word needs to "attend" to all other words in the sequence. This happens through three key steps using learned weight matrices (W_q, W_k, W_v) to transform the embeddings into queries, keys, and values. When word embeddings are multiplied by these matrices, the result is

```python
import numpy as np

# Input tokens (3 tokens, each with 3 features)
S = np.array([
    [1, 0, 1, 0],   # for "The"
    [0, 1, 1, 1],   # for "cat"
    [1, 1, 0, 1]    # for "sleeps"
])

# Initialize weights for Query, Key, and Value (4x3 matrices)
W_q = np.array(
    [
        [0.2, 0.4, 0.6, 0.8],
        [0.1, 0.3, 0.5, 0.7],
        [0.9, 0.8, 0.7, 0.6],
        [0.5, 0.4, 0.3, 0.2],
    ]
)

W_k = np.array(
```

```
21          [
22              [0.1, 0.3, 0.5, 0.7],
23              [0.6, 0.4, 0.2, 0.1],
24              [0.8, 0.9, 0.7, 0.6],
25              [0.2, 0.1, 0.3, 0.4],
26          ]
27      )
28
29      W_v = np.array(
30          [
31              [0.3, 0.5, 0.7, 0.9],
32              [0.6, 0.4, 0.2, 0.1],
33              [0.8, 0.9, 0.7, 0.6],
34              [0.5, 0.4, 0.3, 0.2],
35          ]
36      )
37
38      # Compute Query, Key, and Value matrices
39      Q = S @ W_q
40      K = S @ W_k
41      V = S @ W_v
```

Next, attention scores are computed by multiplying Q and $K^\top$ and then applying softmax.

```
1      # Compute scaled dot-product attention
2      d = Q.shape[1]   # Feature dimension
3      attention_scores = Q @ K.T / np.sqrt(d)
4
5      def softmax(x):
6          """Compute softmax values for each set of scores in x."""
7          return np.exp(x) / np.sum(np.exp(x), axis=1, keepdims=True
             )
8
9      attention_weights = softmax(attention_scores)
```

The attention weight would be

$$\text{softmax}\left(\frac{Q \cdot K^\top}{\sqrt{4}}\right) = \begin{bmatrix} 0.324 & 0.467 & 0.209 \\ 0.305 & 0.515 & 0.180 \\ 0.346 & 0.432 & 0.222 \end{bmatrix}.$$

The final output captures how each word relates to every other word in the sentence. In this case, "sleeps" pays most attention to "cat" (0.432), some attention to "The" (0.346), and less attention to itself (0.222). Finally, these scores are used to create a weighted sum of the values:

```
1      output = attention_weights @ V
```

The final output is

$$
\text{output} = \begin{bmatrix} 1.536 & 1.519 & 1.265 & 1.157 \\ 1.566 & 1.536 & 1.261 & 1.137 \\ 1.512 & 1.507 & 1.269 & 1.174 \end{bmatrix}.
$$

5.5   Bibliographic Remarks

Transformer architecture has profoundly revolutionized AI and has broad impacts. As classical computing approaches its physical limitations, it is important to ask how we can leverage quantum computers to advance Transformers with better performance and energy efficiency. Besides the fault-tolerant quantum Transformers introduced in Sect. 5.2, multiple works are advancing this frontier from various perspectives.

One aspect is to design novel quantum neural network architectures introduced in Chap. 4 with the intuition from the Transformer, especially the design of the self-attention block. In particular, Li et al. [24] propose the quantum self-attention neural networks and verify their effectiveness with the synthetic datasets of quantum natural language processing. There are several follow-up works along this direction [25–27].

Another research direction is exploring how to utilize quantum processors to advance certain parts of the transformer architecture. Specifically, Gao et al. [28] consider how to compute the self-attention matrix under sparse assumption and show a quadratic quantum speedup. Liu et al. [29] harness quantum neural networks to generate weight parameters for the classical model. In addition, Liu et al. [30] devise a quantum algorithm for the training process of large-scale neural networks, implying an exponential speedup under certain conditions. Several other works consider machine learning related optimization problems [31–34].

Despite the progress made, several important questions remain unresolved. Among them, one key challenge is devising efficient methods to encode classical data or parameters onto quantum computers. Currently, most quantum algorithms can only implement one or at most constant layers of Transformers [20, 35, 36] without quantum tomography. Are there effective methods that can be generalized to multiple layers, or is achieving this even necessary? Moreover, given the numerous variants of the classical Transformer architecture, can these variants also benefit from the capabilities of quantum computers? Lastly, if one considers training a model directly on a quantum computer, is it possible to do so in a "quantum-native" manner—avoiding excessive data read-in and read-out operations?

References

1. Vaswani, A. (2017). Attention is all you need. In *Advances in neural information processing systems.*
2. Sennrich, R., Haddow, B., & Birch, A. (2016). Neural machine translation of rare words with subword units. In K. Erk & N. A. Smith (Eds.), *Proceedings of the 54th Annual Meeting of the Association for Computational Linguistics, Berlin, Germany, August 2016* (Vol. 1: Long papers, pp. 1715–1725). Association for Computational Linguistics. https://doi.org/10.18653/v1/P16-1162. https://aclanthology.org/P16-1162
3. Kudo, T., & Richardson, J. (2018). SentencePiece: A simple and language independent subword tokenizer and detokenizer for neural text processing. In E. Blanco & W. Lu (Eds.), *Proceedings of the 2018 Conference on Empirical Methods in Natural Language Processing: System Demonstrations*, Brussels, Belgium, November 2018 (pp. 66–71). Association for Computational Linguistics. https://doi.org/10.18653/v1/D18-2012. https://aclanthology.org/D18-2012.
4. Mielke, S. J., Alyafeai, Z., Salesky, E., Raffel, C., Dey, M., Gallé, M., Raja, A., Si, C., Lee, W. Y., Sagot, B., & Tan, S. (2021). Between words and characters: A brief history of open-vocabulary modeling and tokenization in NLP. arXiv:2112.10508.
5. He, K., Zhang, X., Ren, S., & Sun, J. (2015). Deep residual learning for image recognition. arXiv:1512.03385.
6. Ba, J. L., Kiros, J. R., & Hinton, G. E. (2016). Layer normalization. https://arxiv.org/abs/1607.06450
7. Kingma, D. P., & Ba, J. (2015). Adam: A method for stochastic optimization. In Y. Bengio & Y. LeCun (Eds.), *3rd International Conference on Learning Representations, ICLR 2015, San Diego, CA, May 7–9, 2015, Conference Track Proceedings.* http://arxiv.org/abs/1412.6980
8. Jacob, B., Kligys, S., Chen, B., Zhu, M., Tang, M., Howard, A., Steiner, B., & Rolland, H. (2018). Quantization and training of neural networks for efficient integer-arithmetic-only inference. In *Proceedings of the IEEE Conference on Computer Vision and Pattern Recognition (CVPR)* (pp. 2704–2713).
9. Han, S., Pool, J., Tran, J., & Dally, W. J. (2015). Learning both weights and connections for efficient neural networks. In *Advances in neural information processing systems (NeurIPS)* (pp. 1135–1143).
10. Shoeybi, M., Patwary, M., Puri, R., LeGresley, P., Casper, J., & Catanzaro, B. (2019). Megatron-lm: Training multi-billion parameter language models using model parallelism. arXiv preprint arXiv:1909.08053.
11. McDonald, J., Li, B., Frey, N., Tiwari, D., Gadepally, V., & Samsi, S. (2022). Great power, great responsibility: Recommendations for reducing energy for training language models. In *Findings of the Association for Computational Linguistics: NAACL 2022.* https://doi.org/10.18653/v1/2022.findings-naacl.151
12. Desislavov, R., Martínez-Plumed, F., & Hernández-Orallo, J. (2023). Trends in AI inference energy consumption: Beyond the performance-vs-parameter laws of deep learning. *Sustainable Computing: Informatics and Systems, 38*, 100857. ISSN 2210-5379. https://doi.org/10.1016/j.suscom.2023.100857. https://www.sciencedirect.com/science/article/pii/S2210537923000124
13. Yang, G., Hu, E. J., Babuschkin, I., Sidor, S., Liu, X., Farhi, D., Ryder, N., Pachocki, J., Chen, W., & Gao, J. (2022). Tensor programs v: Tuning large neural networks via zero-shot hyperparameter transfer. arXiv:2203.03466.
14. Ma, S., Wang, H., Ma, L., Wang, L., Wang, W., Huang, S., Dong, L., Wang, R., Xue, J., & Wei, F. (2024). The era of 1-bit LLMs: All large language models are in 1.58 bits. arXiv:2402.17764.
15. Sun, X., Tian, G., Yang, S., Yuan, P., & Zhang, S. (2023). Asymptotically optimal circuit depth for quantum state preparation and general unitary synthesis. *IEEE Transactions on Computer-Aided Design of Integrated Circuits and Systems, 42*(10), 3301–3314 (2023). ISSN 1937-4151. https://doi.org/10.1109/TCAD.2023.3244885. https://ieeexplore.ieee.org/document/10044235

16. Zhang, X.-M., Li, T., & Yuan, X. (2022). Quantum state preparation with optimal circuit depth: Implementations and applications. *Physical Review Letters, 129*(23), 230504. https://doi.org/10.1103/PhysRevLett.129.230504

17. Low, G. H. (2017). *Quantum signal processing by single-qubit dynamics*. Thesis, Massachusetts Institute of Technology. https://dspace.mit.edu/handle/1721.1/115025

18. Guo, N., Mitarai, K., & Fujii, K. (2024). Nonlinear transformation of complex amplitudes via quantum singular value transformation. *Physical Review Research, 6*, 043227. https://doi.org/10.1103/PhysRevResearch.6.043227

19. Rattew, A. G., & Rebentrost, P. (2023). Non-linear transformations of quantum amplitudes: Exponential improvement, generalization, and applications. https://arxiv.org/abs/2309.09839

20. Guo, N., Yu, Z., Choi, M., Agrawal, A., Nakaji, K., Aspuru-Guzik, A., & Rebentrost, P. (2024). Quantum linear algebra is all you need for transformer architectures. https://arxiv.org/abs/2402.16714

21. Lloyd, S., Mohseni, M., & Rebentrost, P. (2014). Quantum principal component analysis. *Nature Physics, 10*(9), 631–633. ISSN 1745-2481. https://doi.org/10.1038/nphys3029

22. Kerenidis, I., & Prakash, A. (2016). Quantum recommendation systems. arXiv:1603.08675.

23. Hendrycks, D., Burns, C., Basart, S., Zou, A., Mazeika, M., Song, D., & Steinhardt, J. (2021). Measuring massive multitask language understanding. In *Proceedings of the International Conference on Learning Representations (ICLR)*.

24. Li, G., Zhao, X., & Wang, X. (2023). Quantum self-attention neural networks for text classification. https://arxiv.org/abs/2205.05625

25. Cherrat, E. A., Kerenidis, I., Mathur, N., Landman, J., Strahm, M., & Li, Y. Y. (2024). Quantum vision transformers. *Quantum, 8*, 1265. ISSN 2521-327X. https://doi.org/10.22331/q-2024-02-22-1265

26. Evans, E. N., Cook, M., Bradshaw, Z. P., & LaBorde, M. L. (2024). Learning with sasquatch: A novel variational quantum transformer architecture with Kernel-based self-attention. https://arxiv.org/abs/2403.14753

27. Widdows, D., Aboumrad, W., Kim, D., Ray, S., & Mei, J. (2024). Quantum natural language processing. https://arxiv.org/abs/2403.19758

28. Gao, Y., Song, Z., Yang, X., & Zhang, R. (2023). Fast quantum algorithm for attention computation. https://arxiv.org/abs/2307.08045

29. Liu, C.-Y., Yang, C.-H. H., Hsieh, M.-H., & Goan, H.-S. (2024). A quantum circuit-based compression perspective for parameter-efficient learning. https://arxiv.org/abs/2410.09846

30. Liu, J., Liu, M., Liu, J.-P., Ye, Z., Wang, Y., Alexeev, Y., Eisert, J., & Jiang, L. (2024). Towards provably efficient quantum algorithms for large-scale machine-learning models. *Nature Communications, 15*(1), 434.

31. Yang, S., Guo, N., Santha, M., & Rebentrost, P. (2023). Quantum Alphatron: Quantum advantage for learning with Kernels and noise. *Quantum, 7*, 1174. ISSN 2521-327X. https://doi.org/10.22331/q-2023-11-08-1174

32. Zhang, C., & Li, T. (2024). Comparisons are all you need for optimizing smooth functions. https://arxiv.org/abs/2405.11454

33. Wang, H., Zhang, C., & Li, T. (2024). Near-optimal quantum algorithm for minimizing the maximal loss. https://arxiv.org/abs/2402.12745

34. Rebentrost, P., Schuld, M., Wossnig, L., Petruccione, F., & Lloyd, S. (2018). Quantum gradient descent and newton's method for constrained polynomial optimization. https://arxiv.org/abs/1612.01789

35. Liao, Y., & Ferrie, C. (2024). GPT on a quantum computer. https://arxiv.org/abs/2403.09418

36. Khatri, N., Matos, G., Coopmans, L., & Clark, S. (2024). Quixer: A quantum transformer model. https://arxiv.org/abs/2406.04305

Chapter 6
Conclusion

Abstract This chapter concludes the tutorial by highlighting the potential of quantum machine learning (QML) to accelerate scientific discovery and real-world applications. It revisits the tutorial's key themes, including quantum adaptations of classical models, theoretical insights, and implementation on near-term and future quantum devices. The chapter also emphasizes the importance of QML in domains such as drug discovery, material science, and optimization, where classical methods face scalability limits.

This tutorial systematically explores the landscape of quantum machine learning, covering foundational principles, the adaptation of classical models to quantum frameworks, and the theoretical underpinnings of quantum algorithms. By providing practical implementations and discussing emerging research directions, it aims to bridge the gap between classical AI and quantum computing for researchers and practitioners.

The insights gained from this tutorial highlight the potential of quantum machine learning to revolutionize various domains, from fundamental scientific research to practical applications in industry. As quantum hardware continues to evolve, quantum machine learning is likely to play a central role in harnessing the computational advantages of quantum systems. In the meantime, the field of quantum machine learning faces challenges, as discussed at the end of each chapter. Overcoming these barriers will be critical for unlocking its full potential.

Moving forward, interdisciplinary collaboration between quantum computing and AI researchers will be essential for addressing these challenges and realizing the transformative potential of QML. Overall, it is hoped that this tutorial serves as a valuable resource for those eager to contribute to this rapidly evolving discipline.

The insights presented in this tutorial underscore the transformative potential of QML, spanning fundamental scientific research to practical industrial applications. While this tutorial primarily focuses on machine learning, QML also holds promise in several key domains. In drug discovery, QML can enhance molecular interaction simulations, accelerating the identification of promising drug candidates and reducing development costs. In materials science, quantum models improve the prediction

Y. Du et al., *A Gentle Introduction to Quantum Machine Learning*, https://doi.org/10.1007/978-981-95-1284-3_6

of material properties, facilitating the discovery of superconductors, energy-efficient compounds, and high-performance materials. In optimization, QML algorithms offer advancements in logistics, supply chain management, and energy distribution, improving efficiency across industries.

As quantum hardware advances, QML is expected to play a central role in harnessing quantum computational advantages. However, significant challenges remain, as discussed at the end of each chapter. Overcoming these obstacles is crucial for unlocking the full potential of QML. Moving forward, interdisciplinary collaboration between quantum computing and AI researchers will be key to addressing these challenges and driving future innovations.

Ultimately, this tutorial is hoped to serve as a valuable resource for those eager to contribute to this rapidly evolving field.

Appendix A
Concentration Inequality

In this section, we introduce some of the most common concentration inequalities in statistical learning theory. These inequalities are widely used in deriving generalization error bounds for learning models. In practical scenarios, one often needs to infer properties of an unknown distribution based on finite data samples drawn from that distribution. Concentration inequalities address the deviations of functions of independent random variables from their expectations. They provide tools to analyze the difference between the empirical mean (or some estimate) and the true expectation of random variables that follow a probability distribution.

We begin by recalling some basic tools that will be used throughout this section. For any nonnegative random variable X following the probability distribution $p(x)$, its expectation can be written as

$$\mathbb{E}X = \int_0^\infty xp(x)\mathrm{d}x. \tag{A.1}$$

This leads directly to a fundamental building block for concentration inequalities, namely, Markov's inequality.

Lemma A.1 (Markov's Inequality) *For any nonnegative random variable X, and a positive constant $t > 0$, we have*

$$\mathbb{P}\{X \geq t\} \leq \frac{\mathbb{E}X}{t}. \tag{A.2}$$

Y. Du et al., *A Gentle Introduction to Quantum Machine Learning*,
https://doi.org/10.1007/978-981-95-1284-3

Proof Employing the definition of cumulative distribution function, we have

$$\mathbb{P}\{X \geq t\} = \int_t^\infty p(x)\mathrm{d}t$$

$$\leq \int_t^\infty p(x)\frac{x}{t}\mathrm{d}t$$

$$\leq \frac{\int_0^\infty p(x)x\mathrm{d}}{t}$$

$$\leq \frac{\mathbb{E}X}{t},$$
(A.3)

where the first inequality follows that $x/t > 1$ in the interval $x \in [t, \infty]$ and the second inequality employs the positiveness of integral term $p(x)x/t$. ☐

Using Markov's inequality, it follows that if ϕ is a strictly monotonically increasing, nonnegative function, then for any random variable X and real number $t > 0$, we have

$$\mathbb{P}\{X \geq t\} = \mathbb{P}\{\phi(X) \geq \phi(t)\} \leq \frac{\mathbb{E}\phi(X)}{\phi(t)}.$$
(A.4)

An application of this result with $\phi(x) = x^2$ leads to the simplest concentration inequality, i.e., Chebyshev's inequality.

Lemma A.2 *Let X be an arbitrary random variable and the real number $t > 0$, then*

$$\mathbb{P}\{|X - \mathbb{E}X| \geq t\} \leq \frac{\mathrm{Var}(X)}{t^2}.$$
(A.5)

Proof Utilizing the extention of Markov's inequality in Eq. (A.4) with setting $\phi(x) = x^2$ yields

$$\mathbb{P}\{|X - \mathbb{E}X| \geq t\} = \mathbb{P}\{|X - \mathbb{E}X|^2 \geq t^2\}$$

$$\leq \frac{\mathbb{E}(X - \mathbb{E}X)^2}{t^2}$$

$$= \frac{\mathrm{Var}(X)}{t^2}.$$
(A.6)

☐

More generally, by taking $\phi(x) = x^q$ ($x \geq 0$), then for any $q > 0$, we obtain the following moment-based inequality:

$$\mathbb{P}\{|X - \mathbb{E}X| \geq t\} \leq \frac{\mathbb{E}(X - \mathbb{E}X)^q}{t^q}. \tag{A.7}$$

Here, the parameter q can be chosen to optimize the upper bound in specific examples. Such moment bounds often provide sharp estimates for tail probabilities. A related idea forms the basis of Chernoff's bounding method. In particular, by setting $\phi(x) = e^{sx}$ for some $s > 0$, we can derive a useful upper bound for any random variable X and $t > 0$:

$$\mathbb{P}\{X \geq t\} = \mathbb{P}\{e^{sX} \geq e^{st}\} \leq \frac{\mathbb{E}e^{sX}}{e^{st}}. \tag{A.8}$$

In Chernoff's method, the goal is to choose an appropriate $s > 0$ to minimize the upper bound or make it as small as possible.

Now, we turn to concentration inequalities for sums of independent random variables. Specifically, we aim to bound probabilities of deviations from the mean, i.e., $\mathbb{P}\{|S_n - \mathbb{E}S_n| \geq t\}$, where $S_n = \sum_{i=1}^{n} X_i$, and $X_1, \cdots, X_n$ are independent real-valued random variables.

By applying Chebyshev's inequality to S_n, we obtain

$$\mathbb{P}\{|S_n - \mathbb{E}S_n| \geq t\} \leq \frac{\mathrm{Var}(S_n)}{t^2} = \frac{\sum_{i=1}^{n} \mathrm{Var}(X_i)}{t^2}. \tag{A.9}$$

In terms of the sample mean, this can be rewritten as

$$\mathbb{P}\left\{\left|\frac{1}{n}\sum_{i=1}^{n} X_i - \mathbb{E}X_i\right| \geq \varepsilon\right\} \leq \frac{\sigma^2}{n\varepsilon^2}, \tag{A.10}$$

where $\sigma^2 = \frac{1}{n}\sum_{i=1}^{n} \mathrm{Var}(X_i)$. Chernoff's bounding method is particularly useful for bounding tail probabilities of sums of independent random variables. By exploiting the independence property (i.e., the expected value of a product of independent random variables equals the product of their expected values), Chernoff's bound can be expressed as

$$\mathbb{P}\{S_n - \mathbb{E}S_n \geq t\} \leq e^{-st}\mathbb{E}\left[\exp\left(s\sum_{i=1}^{n}(X_i - \mathbb{E}X_i)\right)\right]$$

$$= e^{-st}\prod_{i=1}^{n}\mathbb{E}\left[\exp\left(s(X_i - \mathbb{E}X_i)\right)\right] \quad \text{(by independence)}. \tag{A.11}$$

Now, the challenge then becomes finding a good upper bound for the moment generating function of the random variables $X_i - \mathbb{E}X_i$. For bounded random variables, one of the most elegant results is Hoeffding's inequality [6].

Lemma A.3 (Hoeffding's Inequality) *Let X be a random variable with $\mathbb{E}X = 0$, $a \leq X \leq b$. Then, for $s > 0$*

$$\mathbb{E}[e^{sx}] \leq \exp\left(\frac{s^2(b-a)^2}{8}\right) \tag{A.12}$$

This lemma, combined with Eq. (A.11), immediately implies Hoeffding's tail inequality [6].

Theorem A.1 *Let $X_1, \cdots, X_n$ be independent bounded random variables such that X_i falls in the interval $[a_i, b_i]$ with probability one. Then, for any real number $t > 0$, we have*

$$\mathbb{P}\{S_n - \mathbb{E}S_n \geq t\} \leq \exp\left(\frac{-2t^2}{\sum_{i=1}^{n}(b_i - a_i)^2}\right), \tag{A.13}$$

and

$$\mathbb{P}\{S_n - \mathbb{E}S_n \leq -t\} \leq \exp\left(\frac{-2t^2}{\sum_{i=1}^{n}(b_i - a_i)^2}\right), \tag{A.14}$$

Hoeffding's inequality, first proven for binomial random variables by [2] and [9], provides a powerful tool for bounding tail probabilities. However, a limitation is that it does not take into account the variance of the X_i's, which can sometimes yield loose bounds.

Appendix B
Haar Measure and Unitary t-Design

In this section, we introduce some basic knowledge of Haar measure [4] and unitary t-design [3], which are extensively employed in group representation theory and quantum information [1, 8], especially in the analysis of barren plateaus and the trainability of variational quantum algorithms [7].

We begin with the Haar measure. Roughly speaking, Haar measure is a unique probability measure that generates the uniform distribution over a compact group. In this chapter, we focus on the Haar measure on the unitary space $\mathcal{U}(d)$ for convenience. Mathematically, the Haar measure is uniform given by invariant properties in Definition B.1.

Definition B.1 (Haar Measure on $\mathcal{U}(d)$) A measure μ is the Haar measure on the unitary space $\mathcal{U}(d)$ if and only if μ is:

1. Left invariant, i.e., $\mu(U\mathcal{S}) = \mu(\mathcal{S})$ for any measurable set $\mathcal{S} \subseteq \mathcal{U}(d)$ and any unitary $U \in \mathcal{U}(d)$.
2. Right invariant, i.e., $\mu(\mathcal{S}U) = \mu(\mathcal{S})$ for any measurable set $\mathcal{S} \subseteq \mathcal{U}(d)$ and any unitary $U \in \mathcal{U}(d)$.
3. A probability measure, i.e., $\int d\mu(U) = 1$.

As provided in Definition B.1, the Haar measure is continuous on the whole space due to left- and right-invariant properties. Therefore, researchers are interested in approximating the Haar measure with the uniform distribution over some finite sets. Specifically, the unitary set whose uniform distribution shares the same t-th moment with the Haar measure is defined as the unitary t-design.

Definition B.2 (Unitary 2-Design) Let μ be the Haar measure on the space $\mathcal{U}(d)$. Then, a finite set $\mathcal{S}$ forms a unitary t-design if and only if it fulfills one of the following equivalent conditions:

© The Author(s), under exclusive license to Springer Nature Singapore Pte Ltd. 2025
Y. Du et al., *A Gentle Introduction to Quantum Machine Learning*,
https://doi.org/10.1007/978-981-95-1284-3

1.

$$\frac{1}{|\mathcal{S}|} \sum_{U \in \mathcal{S}} U^{\otimes t} \otimes (U^{\dagger})^{\otimes t} = \int_{\mathcal{U}(d)} U^{\otimes t} \otimes (U^{\dagger})^{\otimes t} d\mu(U).$$

2. Let $P_{t,t}(U)$ be the polynomial with at most t degrees of elements from U and at most t degrees of elements from $U^{\dagger}$. Then

$$\frac{1}{|\mathcal{S}|} \sum_{U \in \mathcal{S}} P_{t,t}(U) = \int_{\mathcal{U}(d)} P_{t,t}(U) d\mu(U).$$

Corollary B.1 *A unitary t-design is also a unitary $(t-1)$-design.*

We remark that invariant properties of the Haar measure lead to several useful formulations of unitary t-designs provided in Facts B.1 and B.2.

Fact B.1 (Average over Unitary 1-Design [10]) *Let $\mathcal{S}$ be a set of unitary 1-design on $\mathcal{U}(d)$ and μ be the corresponding Haar measure. Then*

$$\frac{1}{|\mathcal{S}|} \sum_{U \in \mathcal{S}} U_{ij} U_{i'j'}^* = \int_{\mathcal{U}(d)} U_{ij} U_{i'j'}^* d\mu(U) = \frac{1}{d} \delta_{ii'} \delta_{jj'}.$$

Fact B.2 (Average over Unitary 2-Design [10]) *Let $\mathcal{S}$ be a set of unitary 2-design on $\mathcal{U}(d)$ and μ be the corresponding Haar measure. Then*

$$\frac{1}{|\mathcal{S}|} \sum_{U \in \mathcal{S}} U_{i_1 j_1} U_{i_2 j_2} U_{i_1' j_1'}^* U_{i_2' j_2'}^* = \int_{\mathcal{U}(d)} U_{i_1 j_1} U_{i_2 j_2} U_{i_1' j_1'}^* U_{i_2' j_2'}^* d\mu(U)$$

$$= \frac{1}{d^2 - 1} \left(\delta_{i_1 i_1'} \delta_{i_2 i_2'} \delta_{j_1 j_1'} \delta_{j_2 j_2'} + \delta_{i_1 i_2'} \delta_{i_2 i_1'} \delta_{j_1 j_2'} \delta_{j_2 j_1'} \right)$$

$$- \frac{1}{d(d^2 - 1)} \left(\delta_{i_1 i_1'} \delta_{i_2 i_2'} \delta_{j_1 j_2'} \delta_{j_2 j_1'} + \delta_{i_1 i_2'} \delta_{i_2 i_1'} \delta_{j_1 j_1'} \delta_{j_2 j_2'} \right).$$

How big is a unitary t-design? [11] have proved that, for instance, the size of unitary 1-design and unitary 2-design scale polynomially to the dimension of the unitary space.

Fact B.3 (The Size of a Unitary 2-Design [11]) *A unitary 1-design on $\mathcal{U}(d)$ has no fewer than d^2 elements. A unitary 2-design on $\mathcal{U}(d)$ has no fewer than $d^4 - 2d^2 + 2$ elements.*

For a system with N qubits, the dimension of the unitary space is $d = 2^N$. Therefore, an exact t-design could involve exponential numbers of ensembles with increased qubits. Could we obtain an approximation to the unitary t-design, which can be generated in polynomial times with less degree of freedom? [5] has proved that random quantum circuits with linear depths could form an approximate unitary t-design.

Definition B.3 (Approximate Unitary Designs) Let μ be the Haar measure on the space $\mathcal{U}(d)$. We denote the moment superoperator $\Phi_\nu^{(t)}(A) := \int_{\mathcal{U}(d)} U^{\otimes t} A (U^\dagger)^{\otimes t}$. Denote by $M_n(C)$ the $n \times n$ complex matrices. Denote by $\|\Phi\|_\diamond := \max_{X; \|X\|_1 \le 1} \|(\Phi \otimes \mathcal{I}_n) X\|_1$ the diamond norm for the linear transformation $\Phi : M_n(C) \to M_m(C)$ and $X \in M_{n^2}(C)$. Then, a probability distribution ν on $\mathcal{U}(d)$ is an ϵ-approximate unitary t-design if

$$\left\| \Phi_\nu^{(t)} - \Phi_\mu^{(t)} \right\|_\diamond \le \frac{\epsilon}{d^t}.$$

Fact B.4 (Random Quantum Circuits Form Approximate Unitary Designs, Informal Version from [5]) *For the number of qubits $N \ge O(\log t)$, alternative layered random quantum circuits with Haar-random unitary gates sampled from $\mathcal{U}(4)$ lead to an ϵ-approximate unitary t-design when the circuit depth*

$$k \ge O\left(t^{4+o(1)} \left(Nt + \log \frac{1}{\epsilon} \right) \right),$$

where the term $o(1) \to 0$ when $t \to \infty$.

References

1. Adam, M. (2013). *Applications of unitary k-designs in quantum information processing*. PhD thesis, Masarykova univerzita, Fakulta informatiky.
2. Chernoff, H. (1952). A measure of asymptotic efficiency for tests of a hypothesis based on the sum of observations. *Annals of Mathematical Statistics, 23*, 493–507.
3. Dankert, C., Cleve, R., Emerson, J., & Livine, E. (2009). Exact and approximate unitary 2-designs and their application to fidelity estimation. *Physical Review A Atomic, Molecular, and Optical Physics, 8*(1), 012304.
4. Haar, A. (1933). Der massbegriff in der theorie der kontinuierlichen gruppen. *Annals of Mathematics, 34*(1), 147–169.
5. Haferkamp, J. (2022). Random quantum circuits are approximate unitary t-designs in depth $o(nt^{5+o(1)})$. *Quantum, 6*, 795.
6. Hoeffding, W. (1994). Probability inequalities for sums of bounded random variables. In: Fisher, N.I., Sen, P.K. (Eds.), *The collected works of wassily hoeffding*. Springer Series in Statistics. Springer.
7. Larocca, M., Thanasilp, S., Wang, S., Sharma, K., Biamonte, J., Coles, P. J., Cincio, L., McClean, J. R., Holmes, Z., & Cerezo, M. (2024). A review of barren plateaus in variational quantum computing. arXiv preprint arXiv:2405.00781.
8. Nagy, G. (1993). On the haar measure of the quantum su (n) group. *Communications in Mathematical Physics, 153*, 217–217.
9. Okamoto, M. (1959). Some inequalities relating to the partial sum of binomial probabilities. *Annals of the Institute of Statistical Mathematics, 10*, 29–35.

10. Puchała, Z., & Miszczak, J. A. (2017). Symbolic integration with respect to the haar measure on the unitary groups. *Bulletin of the Polish Academy of Sciences. Technical Sciences, 65*(1), 21–27.
11. Roy, A., & Scott, A. J. (2009). Unitary designs and codes. *Designs, Codes and Cryptography, 53*, 13–31.